José Luis Rios Flores
Sigifredo Armendáriz Erives
Gonzalo Hernández Ibarra

Physical, economic and social productivity of used water

José Luis Rios Flores
Sigifredo Armendáriz Erives
Gonzalo Hernández Ibarra

Physical, economic and social productivity of used water

Gravity irrigated date (Phoenix dactylifera L.) production in Comondú, Baja California Sur

ScienciaScripts

Imprint

Any brand names and product names mentioned in this book are subject to trademark, brand or patent protection and are trademarks or registered trademarks of their respective holders. The use of brand names, product names, common names, trade names, product descriptions etc. even without a particular marking in this work is in no way to be construed to mean that such names may be regarded as unrestricted in respect of trademark and brand protection legislation and could thus be used by anyone.

Cover image: www.ingimage.com

This book is a translation from the original published under ISBN 978-3-659-65264-6.

Publisher:
Sciencia Scripts
is a trademark of
Dodo Books Indian Ocean Ltd. and OmniScriptum S.R.L publishing group

120 High Road, East Finchley, London, N2 9ED, United Kingdom
Str. Armeneasca 28/1, office 1, Chisinau MD-2012, Republic of Moldova, Europe
Printed at: see last page
ISBN: 978-620-7-67272-1

Contents

SUMMARY

The objective of this work was to determine the economic-social productivity of water used in the production of date palm (*Phoenix dactylifera L.*) in the DDR (Rural Development District) Comondu, Baja California Sur, Mexico and to contrast it against the corresponding indicators of grain maize (*Zea mays*) cultivation in the same DDR, through indicators of physical (PFA), economic (PEA) and social (PSA) productivity of water. In the methodology, the mathematical models of Rios *et al.* (2015 and 2018) were used to estimate the PFA, PEA and PSA. The results show that the PFA, PEA and PSA of date and grain maize crops were: in the PFA: 0.103 and 1.550 kg m^{-3} , in the PEA: USD 0.083 and USD 0.009 profit per m^3 and in the PSA: 3.84 and 4.09 jobs hm^{-3} respectively. The EAP of dates was higher than crops such as maize grain from the same DDR, rainfed coffee from Chiapas, walnuts from Chihuahua and Coahuila, and bovine milk from Chihuahua, but lower than apples with low, medium or high use of technology from Chihuahua, grapes from Sonora and Coahuila, avocado from Michoacan, peaches from Zacatecas. The PES of date and maize grain had a PES index of 3.84 and 4.09 hm^{-3} , the PES of date exceeded only the PES of Comondu orange and Zacatecas peach, but was lower than the PES of Chihuahua apple, Sonora and Coahuila grapevine, Chihuahua and Coahuila walnut, Michoacan avocado and Chiapas coffee.

Keywords: Water Efficiency, Virtual Water, Sustainability, Water Footprint

I. INTRODUCTION

In terms of territorial extension, the water on Earth represents 70%, only 30% is the surface of the Earth itself^ hence our planet should not really be called the planet Earth, but the planet "Water", although, with a certain sense of justice towards water, the planet is often also called "The blue planet", hence common sense tricks us into believing that the water resource is unlimited, almost infinite, so we should not have to worry about using water in the most efficient, productive way (see box 1), why worry about water if it is abundant?. Nothing could be more false than this, because, according to the "Aqua Foundation" "our blue planet contains some 1386 million km^3 of water, an amount that has neither decreased nor increased in the last two million years.

It is estimated that 97.5% is salt water and only 2.5% of the Earth's water is considered fresh. If we take into account that 90% of the planet's available freshwater resources are in the Antarctic, this sense of abundance is diminished. Only 0.5% of fresh water is found in underground reservoirs and 0.01% in rivers and lakes.

So how much drinkable water is there on Earth? Official figures state that only 0.007% of the Earth's water is drinkable, and that amount is decreasing year by year due to pollution. This makes us aware that water is a scarce and limited resource and a right in an unequal world. Lack of access to water is a cause of poverty, inequality, social injustice and creates huge differences in life chances. The UN confirms that water scarcity affects more than 40% of the world's population. Every day, nearly a thousand children die from preventable water-borne diseases or sanitation-related diarrhoea. That is why ensuring access to safe drinking water is one of the Millennium Development Goals. Water is the source of life.[1]"

Others go further, and give figures that numerically characterise the scarcity of water, specifically of available freshwater, see table 1.

From the above two sources it is clear that the available freshwater on the planet is indeed scarce, as much of that freshwater is not available to humans, as more than 8,400,000 km[1 2 3] of the earth's freshwater is more than one km deep, the majority of the freshwater, 29,200,000 km^3 is found mainly in polar regions and Greenland, as well as in the permafrost not only of Greenland, but also of Siberia and North America.

Box 1. ^How much water is there on earth?

Water source	Volume of water km^3	Percentage of freshwater	Percentage of total water
Oceans, seas and bays	1.338.000.000	-	96,54
Polar ice caps, glaciers and permanent snow	24.064.000	68,6	1,74
Groundwater	23.400.000	-	1,69
-Sweet	10.530.000	30,1	0,76
-Salad	12.870.000	-	0,93
Soil moisture	16.500	0,05	0,001

---

[1]Aqua Foundation. Quantity of drinking water, source of life. Available in: https://www.fundacionaquae.org/wiki-aquae/datos-del-agua/cantidad-de-agua-potable-
source-of-life/ Accessed on: 10 September, 2019.
2Cosmo News, 2012. ^How much water is there on earth? Available at: https://www.cosmonoticias.org/cuanta-agua-hay-en-la-tierra/, Accessed on: 10 September, 2019.

Ground ice and permafrost	300.000	0,86	0,022
Lakes	176.400	-	0,013
-Sweet	91.000	0,26	0,007
-Salad	85.400	-	0,007
Atmosphere	12.900	0,04	0,001
Marsh water	11.470	0,03	0,0008
Rfos	2.120	0,006	0,0002
Biological water	1.120	0,003	0,0001

Note: Percentages may not add up to 100% due to rounding.

Source: Cosmo News, 2012. [2]

On the other hand, the number of human beings on the planet has grown exponentially, so that "the Neolithic revolution, ten thousand years ago, through the application of agricultural and livestock techniques, allowed the first great expansion of the human species; it is estimated that from then on the population began to grow at a rate that doubled every seventeen hundred years. At the beginning of our era there were an estimated one hundred and fifty million people: one third in the Roman Empire, another third in the Chinese Empire and the rest scattered"[4] (see figure 1), so, if only 2000 years ago there were 150 million people, and today in 2019 there are 7.7 billion inhabitants[5] , and considering, as stated earlier, that the volume of available water has not changed in the last millennia, then, the *per capita* freshwater endowment available today is equivalent

---

[4] Evolution of the world population. The market economy: virtues and innovations. Demographics. 2019. Available at: http://www.juntadeandalucia.es/averroes/centrostic/14002996/helvia/aula/archivos/repositorio/250/271/html/economia/2/evolucion.htm .

Access date: 10 September, 2019.

[5] BBC News World.2019. World population day: ^how many humans have lived on earth? Available at: https://www.bbc.com/mundo/noticias-48958753. last accessed 10 September, 2019.

to only 1.94%[6] of the *per capita* endowment that was available two thousand years ago.

According to CONAGUA (2015) in 2012 the natural availability in Mexico was 4,028 m^3 per person, and it is estimated that in 2030 it will be only 3,430 m^3 per person, while per capita consumption went from 40 litres per day in 1955 to 280 litres per person per day in 2012[6].

This suggests that available freshwater is indeed an extremely scarce resource, and that population growth is exacerbating it, but, more to the point: ^Which human activity demands the most available freshwater, some sources indicate that agriculture consumes between 69%[7][8] and 70%[9] of the planet's available freshwater, the remaining percentage being consumed by other human activities such as industry (20-21%) and domestic consumption (10%).

Figure 1. Evolution of the world population[9]

---

[6] If "K" is the volume of available freshwater (which some authors put at 41000 km^3), then 2,000 years ago the per capita endowment per million inhabitants was a= K/150, by 2019 it would be b=K/7,700, so (1/7,700)/(1/150)=1.94%.

[7] National Water Commission (2015): *Atlas del Agua en Mexico*. Conagua. Document available

at: http://www.conagua.gob.mx/CONAGUA07/Publicaciones/Publicaciones/ATLAS2015. p df

[8] **FAO, 2002**. Available at: http://www.fao.org/3/y3918s/y3918s03.htm

[9] El Heraldo, Econom^a Section. 22 March 2015. Costa Rica. available at: https://www.elheraldo.co/economia/la-agricultura-consume-el-70-del-agua-en-el-mundo- 188535#.

In the case of Mexico, which has 0.1% of the world's total available freshwater, agriculture and livestock consume 76.3% of available freshwater (the world average is 70%), industry and energy generation consume 13% (the world average is 22%) and domestic consumption 10% (the world average is 8%), but in Mexico, agricultural activity is the one that wastes the most water: 57% of the water it consumes is lost through evaporation, but above all, through inefficient and obsolete irrigation infrastructure, and irrigated agriculture contributes 42% of national production[10][11] .

All of the above indicates that the available freshwater on the planet in general, and in Mexico in particular, is scarce, and that agriculture is the main user, and that it uses it inefficiently, which is why, although agriculture provides food, it should not be exempt from using water efficiently and productively. Therefore, the *problem in which this work is inserted,* is the sustainable use of water given its notorious scarcity, and the *general objective* of this work is not the study of date palm (*VPhoenix dactylifera L.)* or maize grain (*Zea mays*), but to determine the productivity and efficiency with which irrigation water is used, *delimiting it to* the use of water in the cultivation of date in Comondu, Baja California, Mexico, in such a way that the efficiency and productivity of water used in the production of date, will be evaluated from three different perspectives: the

---

[10]Evolution of the world population. The market economy: virtues and innovations. Demographics. available at:

        http://www.juntadeandalucia.es/averroes/centros-tic/14002996/helvia/aula/archivos/repositorio/250/271/html/economia/2/evolucion.htm .

Access date: 10 September, 2019.

[11]Agua.org.mx Fondo para la Comunicacion y la Educacion Ambiental A.C. Overview of water in Mexico. Available at: https://agua.org.mx/cuanta-agua-tiene-mexico/. Last accessed on: 10 September, 2019.

efficiency and **physical** productivity of water, the efficiency and **economic** productivity of water and the efficiency and **social** productivity of water used in the production.

II. PARTICULAR OBJECTIVES AND HYPOTHESES

2.1. Particular objectives

For the cultivation of date palms (*Phoenix dactylifera*) irrigated in the traditional way by gravity-rolled water in the DDR Comondu, Baja California Sur, the particular objectives are to determine indicators of:

a) *Profitability*

b) Physical, economic and social *efficiency* of water use in the production of date palm (*Phoenix dactylifera*).

c) Physical, economic and social *productivity* of water used in the production of date palm (*Phoenix dactylifera*).

2.2. Hypothesis

First hypothesis

The profitability of the cultivation of date (*Phoenix dactylifera*) irrigated in a traditional way by gravity-rolled water in the DDR Comondu, Baja California Sur, Mexico, will have a higher profitability than the cultivation of maize (*Zea mays*) grain, a crop that is one of the main crops in the DDR Comondu, BCS.

Second hypothesis

The cultivation of date (*Phoenix dactylifera*), traditionally irrigated by gravity-rolled water in the DDR Comondu, Baja California Sur, Mexico, in relation to the water used in production, will have a higher ***physical*** water productivity index (PFA, measured in kg m^{-3}) ***higher*** than the PFA index of grain maize (*Zea mays*), i.e. the same volume of water, one m^3 ***, produces more biomass*** in datil (*Phoenix dactylifera)* than in grain maize (*Zea mays)*.

Third hypothesis

The cultivation of date (*Phoenix dactylifera*), traditionally irrigated by gravity-rolled water in the DDR Comondu, Baja

California Sur, Mexico, in relation to the water used in production, will have an *economic* water productivity index (EAP, measured in USD profit per m^3 *) higher* than the EPP index of the grain maize (*Zea mays*) crop, i.e., the same volume of water, one m^3 *, produces more profit* in datil (*Phoenix dactylifera*) than in grain maize (*Zea mays*).

Fourth hypothesis

The cultivation of date palm (*Phoenix dactylifera*), traditionally irrigated by gravity-fed water in the DDR Comondu, Baja California Sur, Mexico, in relation to the water used in production, will have a higher *social* water *productivity* index (WSP), measured in terms of jobs generated per hm) than the WSP index of grain maize (Zea mays), PES measured in jobs generated per hm^3 *)* *higher* than the PES index of the grain maize (*Zea mays*) crop, i.e., the same volume of water, one hm^3 *, produces more jobs* in datil (*Phoenix dactylifera*) than in grain maize (*Zea mays*).

III. LITERATURE REVIEW
III.1. Delimitation of the concept of water productivity and various studies on it
various studies on water productivity

Based on the verbal definitions of the concept of water productivity by Viets (1966)[12] [13] , who restricts it to the physiological aspect of plants, and by Kijne et al (2003)[12] , who delimits the concept of water productivity to the sphere of the capacity of agricultural systems to convert the water used in production into food, but above all by Molden et al (2010)[14] , who defines the concept of water productivity as "The ratio of the net benefits of agricultural, forestry, fisheries, livestock and mixed agricultural systems to the amount of water used to produce those benefits. Fisheries, livestock and mixed farming systems to the amount of water used to produce those benefits. In its broadest sense, it reflects the objectives of producing more food, income, livelihood and ecological benefits at a lower social and environmental cost per unit of water consumed. Physical water productivity is defined as the ratio of agricultural output to the amount of water consumed - "more crop per drop". And economic water productivity is defined as the value derived per unit of water used and this has also been used to relate water used in agriculture to nutrition,

---

[12]**Viets, F.G. 1966.** Increasing water use efficiency by soil management. In Plant environment and efficient water use. Guilford RD, Madison, USA: American Society of Agronomy. Soil science Society of America. 295 p.
[13]**Kijne, J.W., R. Barker and D. Molden, 2003.** Water Productivity in Agriculture: Limits and Opportunity for Improvement. CABI, Cambridge, UK, ISBN: 0 85199 669 8.
[14]**Molden, D; Murray-Rust, H.; Sakthivadiel, R; Makin, I.; 2003.** A water productivity framework for understanding and action, pp.1-18. In: Kijne, J. W.; Barker, R.; Molden, D. J. 2003. Water productivity in agriculture: limits and opportunities for improvement. CABI Publication, Wallingford UK. 332p.

employment, welfare and the environment", Rios et al (2015^{15} a, 2016^{16}, 2018^{17}) reduce the three authors' verbal definitions of water productivity to their mathematical aspect, converting them into the following equations:

$$\mathrm{Pr}\,oductividad = \frac{cantidad\ de\ producto\ "Q"\ (f\acute{\imath}sico, econ\acute{o}mico\ o\ social)}{volumen\,"V"\,de\ agua\ usado\ en\ la\ producci\acute{o}n}$$

$$Eficiencia = \frac{volumen\ "V"\ de\ agua\ usado\ en\ la\ producci\acute{o}n}{cantidad\ de\ producto\ "Q"\ (f\acute{\imath}sico,\ econ\acute{o}mico\ o\ social)}$$

Such equations make it possible to obtain numerous indices of *productivity and efficiency* with which water is used in production, where "Q" can be given in physical (kg, ton), economic (monetary units of income or better still of profit) and social units (expressed in number of working days, jobs), [3]For example, a productivity index would be expressed in kg m-3 (which would indicate how many kg of product are produced for each m3 of water used in production), This would be a physical water productivity index) or in USD m-3 (the index would be suggesting how many USD of income, or profit, as the case may be, are produced for each m3 of water used in production, and thus the water productivity index would be of an economic

---

[15]**Rios-Flores, J. Luis, Torres M., Miriam, Castro F., Rafael, Torres M., M.A. Ruiz T. Jose. 2015 a.** Determination of the blue metric footprint in forage crops of DR-017 Comarca Lagunera, Mexico. Rev. FCA UNCUYO, 2015. 47(1): 101-122, ISSN print 0370-4661. ISSN (online) 1853-8665, pp.93-107. Mendoza, Argentina

[16]**Rios-Flores, Jose Luis, Torres M. M. and Torres M., M. A. (2016 a).** Agricultural water productivity of pecan nut trees in northern Mexico. Cases: Comarca Lagunera and Delicias, Chihuahua. ISBN978-3-639-80166-8. Editorial Academica Espanola. Saarbrucken, Germany.

[17]**Rfos-Flores, Jose Luis, Rios Arredondo, Becky Elizabeth, Cantu Brito, Jesus Enrique, Rios Arredondo, Hebrian Efrain, Armendariz Erives, Sigifredo, Chavez Rivero, Jose Antonio, Navarrete Molina, Cayetano & Castro Franco, Rafael (2018).** Analisis de la eficiencia ffsica, economica y social del agua en esparrago (*Asparagus officinalis L.)* y uva (*Vitis* vinifera) mesa del DR-037 Altar-Pitiquito-Caborca, Sonora, Mexico 2018. *Revista de la Facultad de Ciencias Agrarias. National University of Cuyo,* 50(2). ISSN in print 0370-4661, ISSN (online) 1853-8665. Mendoza, Argentina.

nature) or in jobs hm-3 (the productivity index would be of a social nature, and in this case would indicate how many jobs are associated with the use of one cubic hectometre of water used in production).

By applying the two general models to the particularities of agricultural production, Rios et al (2015, 2016 and 2018 Op. Cit.) determined the particular models of productivity and efficiency of water used in production shown in Table 9 in the materials and methods section of this paper, the models were generated both for an individual crop and for an aggregate of different crops (forage, staple, oilseed, industrial, horticultural or a whole pattern of crops in a region or country).

The water footprint (WH) is a broader concept than water productivity, as WH comprises not only the volume of water used in direct production as in the case of water productivity, but also the virtual water used in the production of inputs and services used in production such as the transport of the product, the concept of HH was originated by Hoekstra (2003)[18] and developed by Hoekstra and Chapagain in 2004[19] , however, it must be stressed, that there are differences between the two concepts, HH apart from the water consumed in the production of the good, as does the concept of water productivity, also comprises the virtual water used in the transportation of that good, as well as the virtual water comprised by the inputs and machinery that were used in the production. Hoekstra and Chapagain

---

[18]**Hoekstra, A.Y. 2003.** Virtual Water Trade: Proceedings of the International Expert Meeting on Virtual Water Trade. Delft. The Netherlands. 12 and 13 December 2002. Value of Water Research Report Series No. 12. UNESCO-IHE. Delft. The Netherlands. www.waterfootprint.org/Reports/Report12.pdf.
[19]**Hoekstra A. Y.; Chapagain A. K. 2004**. Water footprints of Nations. UNESCO-IHE. Institute of Water Education. Value of Water. Research Report Series. Series 16. Volume 1. Netherlands.

differentiate the water footprint into blue (water from surface sources and groundwater), green (rainwater) and grey (water polluted in the production process) and the sum of the three: the total WF.

Table 2 records the summary results found on the physical, economic and social productivity of water used in the production of some fruit crops as well as grains and oilseeds, as well as bovine milk in Mexico, the monetary figures of the EAP (economic productivity of water) are those reported by the authors, indicated in the far right column of the table, and given that they are from different years and are valued in nominal or current figures, it is first necessary to deflate them to constant 2018 USD in order to validate the comparison[20] , Likewise, some of the EAP variables are valued in Mexican pesos (MX\$), others in USD of years other than 2018, and others are expressed in euros, so it is a premise to standardise all the ADP information to a single denomination (constant USD of 2018) and that they all refer to the same volumetric unit of water (one hm^3 , remember that one hm^3 has one million m^3), in this way, once the monetary figures of the ADP were standardised, we have table 3.

Table 3 shows that the average AFP for fruit crops, such as dates, was 1.014 kg m-3 , with extreme levels of AFP, such as apples produced under high technology use (HT) and apples produced under medium technology use (MT) conditions in Chihuahua (with AFP of 3.18 and 2.02 kg m-3 respectively)$^{-3}$, as well as the orange produced there in

---

[20]When monetary figures are valued in nominal terms, i.e. without having previously removed the distorting effect of inflation, it is not valid to make comparisons with them, much less to draw inferences from them, since being valued in nominal terms implies that there is different purchasing power in each of the years to which each of these figures belong.

Comondu, BCS, with an index of 1.65 kg m-3 respectively).18 and 2.02 kg m^{-3} respectively)[21] , as well as the orange[22] produced there in Comondu, BCS, with an index of 1.65 kg m-3, or in crops with a low PFA with indexes of 0.085 kg m3 such as coffee produced under rainfed conditions on average across the state of Chiapas[23] (see tables 2 and 3).

---

[21] **Rios-Flores, J. L., Torres M., M., Azpilcueta Ruiz-Esparza, M. 2017**.Water productivity in apple produced under different levels of technification, in Cuauhtemoc, Chihuahua, Mexico. Rev. Asuntos Economicos y Administrativos No.32, first semester 2017.pp.135-146, Universidad de Manizales, Colombia.

[22] **Cifuentes, G. O. 2013.** Eficiencia ftsica, economica y social del agua irrigada en naranja (*Citrus sinensis*) en Baja California Sur Tesis profesional. Universidad Autonoma Agraria Antonio Narro Unidad Laguna, Torreon, Coahuila.

[23] **Azpilcueta Ruiz-Esparza, M., Rios-Flores, J. L., Ruiz-Torres, J. 2017.** Water productivity in rainfed coffee crops in Villaflores, Chiapas, Mexico. Rev. Asuntos Economicos y Administrativos, first half of 2017. Pp. 183-192. University of Manizales, Colombia.

Table 2. Physical (PFA), economic (PEA) and social (PES) productivity of water used in the production of various perennial and annual agricultural and livestock products. EAP in nominal monetary units.

Product	Place	PFA	Units	EAP	Units	PSA	Units	Author
Orange	Comondu, BCS	1.65	kg пт^3	$ 1.49	MX$ profit per m^3	3.60	Jobs hm^3	Cifuentes, 2013.
apple tree BT	Cuauhtemoc, Chihuahua	0.91	kg пт^3	$84,341	USD profit per hm^3	28.10	Jobs hm^3	Rios *et al (2017)*
Apple AT	Cuauhtemoc, Chihuahua	3.18	kg пт^3	$ 614,244	USD profit per hm^3	22.70	Jobs hm^3	Rios *et al (2017)*
Apple MT	Cuauhtemoc, Chihuahua	2.02	kg пт^3	$ 284,726	USD profit per hm^3	19.60	Jobs hm^3	Rios *et al (2017)*
Temporary coffee	Villafl ores, Chiapas	0.214	kg пт^3	$ 22,992	USD profit per hm^3	26.30	Jobs hm^3	Azpilcueta *et al (2017)*
Temporary coffee	Chiapas	0.085	kg пт^3	-$ 20,295	USD profit per hm^3	16.30	Jobs hm^3	Azpilcueta *et al (2017)*
Vine	Caborca, Sonora	1.6	kg пт^3	$ 945,190	USD profit per hm^3	10.70	Jobs hm^3	Rios *et al (2018)*
Vine	Coahuila	1.09	kg пт^3	$ 223,970	USD profit per hm^3	53.32	Jobs hm^3	Carrillo, 2019
Avocado	Patzcuaro, Michoacan	1.35	kg пт^3	$ 110,000	USD profit per hm^3	31.90	Jobs hm^3	Rios, Torres y Azpilcueta (2019)
Walnut	Delicias, Chihuahua	0.125	kg пт^3	$ 1.01	MX$ profit per m^3	3.90	Jobs hm^3	Rios, Torres y Torres (2016)
Walnut	Southwest of Coahuila	0.064	kg пт^3	$ 29,785	USD profit per hm^3	17.10	Jobs hm^3	Rios, Torres y Rios (2018)
Peach	Fresnillo, Zacatecas	0.5	kg пт^3	$ 1.80	MX$ profit per m^3	2.00	Jobs hm^3	Rios *etal (2015)*

Product	Location	Value	Units	Economic value	Units	Source
Bovine milk	Delicias, Chihuahua	0.213	L пт^3	USD $ 4,500 profit per hm^3		Rios, Rios y Rios (2019)
Bovine milk	China	1,282	m^3 ton^{-1}			Gleick, Peter H. 2004.
Bovine milk	India	1,078	m^3 ton^{-1}			Mekonnen & Hoekstra (2012)
Bovine milk	Netherlands	528	m^3 ton$^{'1}$			Mekonnen & Hoekstra (2012)
Bovine milk	E. U. A.	796	m^3 ton$^{'1}$			Mekonnen & Hoekstra (2012)
Bovine milk	World average	1,020	m^3 ton$^{'1}$			Mekonnen & Hoekstra (2012)
The whole economy	U.S.A.	-		6.8 (1900)-14 (1996)	USD profit per m^3	Gleick, Peter H. 2004
Cotton	Spain	1.04	kg m$^{'3}$	$ 0.23	€ m$^{'3}$	Montesinos *et al* (2011)
Rice	Spain	0.81	kg пт^3	$ 0.42	€ пт^3	Montesinos *et al* (2011)
Strawberry	Spain	0.05	kg m$^{'3}$	$21.40	€ m$^{'3}$	Montesinos *et al* (2011)
Sunflower	Spain	1.23	kg пт^3	$ 0.25	€ пт^3	Montesinos *et al* (2011)
Maize grain	Spain	0.38	kg m$^{'3}$	$ 0.42	€ m$^{'3}$	Montesinos *et al* (2011)
Olive tree	Spain	0.39	kg пт^3	$ 0.97	€ пт^3	Montesinos *et al* (2011)
AVERAGE IN FRUITS		0.945				
GRAND AVERAGE (agricultural products only)		0.927				

Source: Own elaboration. In apple tree: BT, MT y AT are the initials of low technology use, medium technology use and high technology use respectively.

Another fruit crop, grapevine, if produced in Sonora[24] , Mexico, has an AFP of 1,600 kg m^{-3} , and if produced in Coahuila[25] it has an AFP index of 1,090 kg m-3, with an index of 1.350 kg m-3, the avocado produced in Patzcuaro, Michoacan[26] , is the last fruit crop that enjoyed relatively high levels of AFP, since walnut and peach were among the fruit trees with lower AFP, since in the case of pecan walnut from Delicias[27] , Chihuahua, Mexico the AFP was of the order of 0.125 kg m^{-3} , while the walnut produced in the southwest of Coahuila[28] , Mexico, had an index of only 0.064 kg m -3, the peach from Fresnillo, Zacatecas[29] , Mexico was intermediate among the fruit trees with the lowest AFP with an index of 0.500 kg m -3 (see tables 2 and 3).

In Spain, Montesinos *et al* (2011)[29] found that the PFA in agricultural products such as cotton, rice, strawberry, sunflower, grain corn and olive tree, had indexes of 1.040, 0.810, 0.050, 1.230, 0.380 and 0.390 kg m^{-3} respectively (see tables 2 and 3).

The physical Footprint "HHF"[30] [31] in some dairy cattle

[24]Rios *et al (2018 Op. Cit.)*

[25]**Carrillo, C., J. 2019.** Economic-social productivity of water in drip-irrigated grapevine (*Vitis* vinifera) in Coahuila, Mexico. Professional thesis. Universidad Autonoma Chapingo, Bermejillo, Durango, Mexico.

[26]**Rios-Flores, Jose Luis, Ruiz-Torres, J. Azpilcueta Ruiz-Espazra, M. 2019.** Economic-social productivity of water in avocado cultivation. The case of production in Michoacan, Mexico. Editorial Academico Espanola. ISBN 978-3-639-531831. Beau Bassin, Mauritius.

[27]**Rios-Flores, Jose Luis, Torres M., M. Torres M., M. A. 2016 a,** *Op. Cit.*

[28]**Rios-Flores, Jose Luis, Navarrete: Molina, Cayetano, 2017.** Water footprint and economic water productivity in pecan nut (*Carya illinoensis*) in southwestern Coahuila, Mexico. Journal: Studies in Applied Economics International Association of Applied Economics. ISSN 1133-3197, Valladolid, Spain.

[29]**Rios. Flores, J. L., Torres M. M., Ruiz T. J., Torres M. M. A., Cantu B., J. E, 2015 b**. Productive, economic and social evaluation of water in peach (*Prunus persica L Batsch)* in Zacatecas, Mexico. Avances de Investigacion Agropecuaria Journal, University of Colima, Mexico.

[29]Montesinos**, O.; Camacho, E.; Campos, B.; Rodriguez-Diaz, J. 2011.** Analysis of virtual irrigation water. Application to Water Resources Management in a Mediterranean River

products shown in tables 2 and 3 show that the PFA varies from country to country, e.g. according to Mekonnen & Hoekstra (2012)[32] the world average shows that producing one tonne of milk requires 1,020 m^3 of water in production (depending on the density of milk, if you take 1.034 kg per litre of milk would be equivalent to 1,057 litres of water per litre of milk), but if that tonne of milk was produced in China its HH is 1,282 m^3 ton^{-1} , 1,078 m^3 ton^{-1} if it was produced in India, only 528 m^3 ton^{-1} if that milk was produced in the Netherlands, in the U. S. A. the HHF of that tonne of milk is 1,282 m ton , 1,078 m ton if it was produced in India, only 528 m ton if that milk was produced in the U.S. The HHF of milk is 796 m^3 ton^{-1} , and 0.213 litres of milk per m^3 of water used in production in the case of bovine milk from Delicias[33] , Mexico (see tables 2 and 3).

Basin. Water Resources Management. 25 (6): 1635-1651.
[31]Which is used here as a synonym for PFA, but they are not the same, as PFA is only one part of HHF, the other components of HHF are transport, storage, as well as the virtual water consumed by the cattle in their pre-production stage.
[32]**Mekonnen, M.M. & Hoekstra, A. G. (2012).** A global assesment of the water footprint of farm animal products. ECOSYSTEM (2012). 15:401-415. DOI:10.1007-s10021-011-9517-8.
[33]**Rios-Flores, J. Luis; Rios-Arredondo, Becky E.; Rios-Arredondo, Hebrian E. 2019.** Physical and economic footprints of milk. The case of bovine milk from Delicias, Chihuahua, Mexico. Editorial Academica Espanola. ISBN 978-620-0-02518-0. Beau Bassin, Mauritius.

Table 3. Physical (PFA), economic (PEA) and social (PES) productivity of water used in the production of fruit and other agricultural and livestock products. EAP in constant 2018 USD per hm^3 .

Product	Place	PFA Units	Standardised ADP at constant 2018 USD per hm^3	PSA (jobs hm)$^{-3}$	Author
Orange	Comondu, BCS	1.65 kg пт^3	$90,915.53	3.60	Cifuentes, 2013.
apple tree BT	Cuauhtemoc, Chihuahua	0.91 kg пт^3	$90,348.19	28.10	Rios etal (2017)
Apple AT	Cuauhtemoc, Chihuahua	3.18 kg пт^3	$ 657,993.53	22.70	Rios etal (2017)
Apple MT	Cuauhtemoc, Chihuahua	2.02 kg пт^3	$ 305,005.61	19.60	Rios etal (2017)
Temporary coffee	Villaflores, Chiapas	0.214 kg пт^3	$ 24,629.96	26.30	Azpilcueta et al (2017)
Temporary coffee	Chiapas	0.085 kg пт^3	-$21,740.21	16.30	Azpilcueta et al (2017)
Vine	Caborca, Sonora	1.6 kg пт^3	$ 1,012,511.16	10.70	Rios etal (2018)
Vine	Coahuila	1.09 kg пт^3	$ 239,922.26	53.32	Carrillo, 2019
Avocado	Patzcuaro, Michoacan	1.35 kg пт^3	$ 117,834.75	31.90	Rios, Torres y Azpilcueta (2019)
Walnut	Delicias, Chihuahua	0.125 kg пт^3	$ 59,543.29	3.90	Rios, Torres y Torres (2016)
Walnut	Southwest Coahuila	0.064 kg пт^3	$34,178.97	17.10	Rios y Navarrete (2017)

					Rios et al
Peach	Fresnillo,	0.5	kg пт^3	$ 113,674.92 2.00	(2015)
Bovine milk	Delicias, Chihuahua	0.213	L m^{-3}	$ 4,820.51 Sd	Rivers, Rivers and Rivers (2019)
Cotton	Spain	1.04	kg m^{-3}	$ 279,303.77 Sd	Montesinos et al (2011)
Rice	Spain	0.81	kg m^{-3}	$ 510,032.97 Sd	Montesinos et al (2011)
Strawberry	Spain	0.05	kg m^{-3}25,987,394.28	$Sd	Montesinos et al (2011)
Sunflower	Spain	1.23	kg m^{-3}	$ 303,591.05 Sd	Montesinos et al (2011)
Maize grain	Spain	0.38	kg m^{-3}	$ 510,032.97 Sd	Montesinos et al (2011)
Olive tree	Spain	0.39	kg m^{-3} 1,177,933.29	$Sd	Montesinos et al (2011)
AVERAGE FRUIT (without	1.014		kg m^{-3}	$ 300,211.63 19.63	
AVERAGE AGRICULTURE	0.982			$ 1,852,535.66	

Source: Own elaboration, based on table 2.

As for the economic productivity of water "PEA", after deflating the monetary variables reported by the authors in Table 2 and standardising all of them to constant 2018 USD per hm^3 of water used in production, it is shown in Table 3, from that source it is observed that one hm^3 of water used in fruit production (without strawberry and milk) produced an average profit of USD 300,211.63 hm^{-3} , and an average profit of USD1,852,535.66 hm^{-3} on agricultural products in general (without strawberry from Spain and milk in general) (see table 3).

Among the fruit trees, the following stand out for having an EAP (always the EAP in USD of profit per hm^3) higher than the average for fruit production: apple A T with USD 657,993.53, apple MT with USD 305,005.61[34] , grapes from Caborca[35] , Sonora with USD 1,012,511.16 and the olive tree[36] from Spain with USD 1,117,933.29 (see table 3).

While the fruit trees that had an EAP lower than the average fruit EAP were the orange[37] produced there in Comondu, BCS, with USD 90,915.53, the low technology apple tree BT[38] with USD 90,348.19, the seasonal coffee from Chiapas (loss of USD 21,740.21 hm^{-3}) and Villaflores (USD 24,629.96)[38] , grapes produced in Coahuila, Mexico with USD 239,922.[2639] , avocado from Patzcuaro, Michoacan, Mexico with USD 117,834.75[39 40 41] , walnuts produced in Delicias, Chihuahua with USD 59,543.29[42] , the walnut produced in the southwest of Coahuila with USD 34,178.97[43] , and the Zacatecan peach from Fresnillo[44] with USD 113,674.92 hm^{-3} (see table 3).

Of the Spanish crops reported by Montesinos *et al* (2011 *Op. Cit.*), cotton (USD 279,303.77 profit per hm^{-3}) and sunflower (USD 303,591.05 hm) were below the average EAP of fruit trees.05 hm^{-3}), rice, maize grain and Spanish strawberry enjoyed a higher EAP than the average for fruit crops, and even strawberry, higher than the overall average EAP for agricultural products (see table 3).

It is interesting to note that the EAP of bovine milk in

---

[34]Rios *et al* (2017, *Op. Cit.*)
[35]Rios *et al* (2018, *Op. Cit.*)
[36]Montesinos *et al* (2011, *Op. Cit.*)
[37]Cifuentes (2013, *Op. Cit.*)
[38]Rios *et al* (2017, *Op. Cit.*)
[39]Azpilcueta *et al* (2017, *Op. Cit.*)
[40]Carrillo (2019, *Op. Cit.*)
[41]Rios, Torres and Azpilcueta (2019, *Op. Cit.*)
[42]Rios, Torres and Torres (2016, *Op. Cit.*)
[43]Rios and Navarrete (2017, *Op. Cit.*)
[44]Rios *et al (2015 b, p. Cit.)*

Delicias, Chihuahua[45] , with only USD 4,820.51 of profit hm^{-3} , was, after coffee, one of the lowest EAPs (see table 3), but one should be careful, as this does not mean that the dairy sector does not produce profits, no, it is quite the contrary, its profits, at least in Mexico for industrial groups like LALA, are enormous, what makes its EAP very low, is that this dairy sector uses enormous amounts of water.

At a general level, for the entire US economy, for the period 1900 to 1996, Gleick (2002)[45] determines that the EAP was increasing, as one m^3 of water used in the US economy saw the economic productivity of water rise from USD 6.7 to almost USD 14 m^3 (see Figure 2).

In relation to the social productivity of water "PES", table 3 shows that on average, in fruit crops, the use of one hm^3 of water in production is associated with the creation of 19.63 jobs, finding that the crops that were above this average were: apple trees BT, AT with 28.1, 22.7 and apple trees MT[46] [47] were in the average with 19.60 jobs hm^{-3} , seasonal coffee at Villaflores[48] , Chiapas, with 26.30 jobs hm^{-3} , grapes from Coahuila[49] , Mexico, with 53.32 jobs hm^{-3} , avocado from Patzcuaro[50] , Michoacan, Mexico with 31.90 jobs hm^{-3} , while the crops that had a lower than average PES of 19.63 jobs hm^{-3} , were the crops of: orange[51] from Comondu, BCS, seasonal coffee from Chiapas[52] with 16.30

---

[45]**Rios-Flores, J. Luis; Rios-Arredondo, Becky E.; Rios-Arredondo, Hebrian E. 2019.**
Op. cit.
[46]**Gleick, Peter H. 2004.** Global freshwater resources. Soft-Path solutions for the 21st Century.
[47]Rios *et al (2017, Op. Cit.)*
[48]Azpilcueta *et al (2017, Op. Cit.)*
[49]Carrillo (2019, *Op. Cit.*)
[50]Rios, Torres and Azpilcueta (2019, *Op. Cit.*)
[51]Cifuentes (2013, *Op. Cit.*)
[52]Azpilcueta *et al* (2017, *O. Cit.*)

hm jobs^{-3} , grapevine from Caborca[53] , Sonora with 10.70 hm jobs^{-3} , walnut trees from Delicias, Chihuahua, Mexico with 3.90 hm jobs^{-3} [54] , and from the south west of Coahuila[55] with 17.10 hm jobs^{-3} , finally peach[56] produced in Fresnillo, Zacatecas, Mexico with 2.0 hm jobs (see tables 2 and 3).

Figure 2. Economic productivity of water used in the United States of America from 1900 to 1996.

Source: Global Freshwater Resources: Soft-Path Solutions for the 21st Century Peter H. Gleick. https://science.sciencemag.org/content/sci/302/5650/1524.full.pdf

III.2. The production of date palm (*Phoenix dactylifera*)

The world production of dates as reported in the Table 4 shows that in 2001, 5,197,422 tons were produced in the world, with Egypt being the main producer with 21.2% (1,102,350 tons) of the world production, Iran, Saudi Arabia, Iraq, Algeria, United Arab Emirates and Oman, being the main datetil producers in the world with 88.Mexico, although it figured that year in the world production, with its 2,600

---

[53]Rios *et al 2018, Op. Cit.*)
[54]Rios, Torres and Torres (2016, *Op. Cit.*)
[55]Rios and Navarrete (2017, *Op. Cit.*)
[56]Rios *et al* (2015 b, *Op. Cit.*)
[56] **Salomon-Torres, R., Ortiz-Uribe, N., & Villa-Angulo, R. 2017**. The production of date palm (*Phoenix dactylifera L.*) in Mexico. Nueva epoca. Ano 16 No.91, January-June 2017, Ciencias Sociales y Exactas. Autonomous University of Baja California

tons contributed only 0.05% of the world production, but already in 2013, according to figure 3, it increased about 38% to be equal to 7,189,789 tons.

Table 4. Date production by country in 2001.

Country	Tons	Country	Tons
Egypt	1,102,350	Yemen	29,837
Iran	900,000	Mauritania	22,000
Saudi Arabia	712,000	Chad	18,000
Pakistan	550,000	Qatar	16,500
Iraq	400,000	United States	15,875
Algeria	370,000	Kuwait	10,376
United Arab Emirates	318,000	Israel	9,484
Oman	260,000	Turkey	9,400
Sudan	177,000	Somalia	10,000
China	110,000	Niger	7,600
Tunisia	107,000	Spain	7,000
Morocco	32,400	Mexico	2,600
Total			5,197,422

Source: Info Agro. Date cultivation (part). Based on FAO figures, 2001. Available at:

https://www.infoagro.com/frutas/frutas_tropicales/datil.htm

Viewed by continent, according to Salomon, Ortiz and Villa (2017)[56] in 2013, the top ten datetil producing countries, Egypt, Iran, Saudi Arabia, Algeria, Iraq, Pakistan, Oman, United Arab Emirates, Tunisia and Libya, produced 91.6% of world datetil production, while America and Europe contributed 0.4% and 0.2% respectively.

At the level of Mexico alone, Figure 3 shows that in 2015 the production of date in Mexico was characterised by 1,052.5 ha harvested, with a production of 7,427.1 tons, production that had a market value of MX$ 326,239.35 thousand.

State	Municipality	Superfine seeded	Superfine harvested (Hi)	Production (Tons)	Yield (Ton/Ha)	Average Price Rural (.(.Ton)	Production Value (Thousands of pesos)
Bai a California	Mexicali	668.25	303.00	2411.84	7.96	57,826.65	139,468.62
Baia Southern California	ComondLi	112.00	121.00	128.26	1.06	40,000.00	5,130.40
Baja California	La Paz	22.50	0.00	0.00	0.00	0.00	0.00

South							
Bai a California Snr	Mnlege	204.50	14.50	29.00	2.00	35 627.59	1 033.20
Coahuila	Viesca	15.00	5.00	12.60	2.52	38000.00	478.80
Sonora	Allar	2.00	2.00	10.40	5.20	39 500.00	410.80
Sonora	Caborca	7.00	7.00	35.00	5.00	40 500.00	1417.50
Sonora	San Luis Rio Colorado	900.00	600.00	4800.00	8.00	37 145.84	178 300.02
Totals		1940.25	1052.50	7427.10	7.06	43 925.54	326 239.35

Figure 3. Production of date palm (*Phoenix dactylifera*) in Mexico in 2015. Figures from SIAP-SAGARPA (2016). Cited by Salomon, Ortiz and Villa (2017 Op. Cit).

Figure 3 shows that in terms of physical yield, the national average was 7.06 tons of date per ha, ranging from 1.06 tons per ha^{-1} in Comondu, BCS, to 8.0 tons per ha^{-1} in San Luis Rio Colorado, Sonora, according to Figure 1, Sonora is the main producer, with 4,845.4 tons, representing 65.24% of the production.

The second largest producer was Baja California with 2,540.1 tons, equivalent to 34.2% of the national production, while Baja California Sur and Coahuila together contributed the remaining 0.56%.

III.3. Characteristics of the date palm (*Phoenix dactylifera* L.)

The date palm (*Phoenix dactylifera L.*) is classified as follows according to Dransfield and Uhi (1986[57], cited by Zaid, 2002):[58] [59]

Group: Spadicifora

Order: Palmea

Family: Palmaceae

Subfamily: Coryphyoideae

Tribe: Phoeniceae

Gender: Fenix

Species: Dactylifera L.

However, the Wikipedia source, which classifies it as it appears in figure 4, coincides with the botanical description of Zaid (2002 *Op. Cit.*) in which the description of Dransfield and Uhi (1986) is cited.

---

[58] **DRANSFIELD, J. and NW UHL. (1986):** A summary of a classification of palms. Principles 30 (1): 3-11. Cited by **Abdelouahhab, Zaid, 2002.**

[59]**Abdelouahhab, Zaid, 2002.** Date palm cultivation. FAO PLANT PRODUCTION AND PROTECTION PAPER. 156 REV.1 ISSN 0259-2517 ISBN 92-5-104863-0. Rome, Italy. Available at: http://www.fao.org/3/y4360e/y4360e00.htm

Figure 4. Taxonomy of the date palm (*Phoenix dactylifera L.*).

The origin of the date palm (*Phoenix dactylifera L.*) is either from North Africa or Arabia[60] , although other sources place its phylogenetic origin in Mesopotamia, between the Tigris River and the Euphrates River, in what would be modern-day Iraq, it was cultivated 50,000 years ago, and from there it spread to the Afro-Asian countries with dry climates, from Morocco to Pakistan, It was introduced to America by Spanish missionaries, and is still cultivated in California, Arizona, Texas and Mexico, among the last countries that have been incorporated to its cultivation are South Africa, Australia, Greece, Sicily, and southern Europe. It is a

---

[60]**Infoagro.** Date cultivation (part 1):
https://www.infoagro.com/frutas/frutas_tropicales/datil.htm

dioecious, slender tree that can reach 25-30 m in height and 2 m in diameter at the base of the stem. Stem: stout, straight, inert, unbranched, covered by the bases of dead leaves, crowned at the apex by a tuft of living leaves. The lower part usually bears numerous adventitious roots, which give rise to suckers, particularly when the palm is still young, so multiple plants can develop if not pruned[60] .

The botanical name of the date palm, *Phoenix dactylifera L.,* according to Abdelouahhab (2002 *Op. Cit.*) comes from the Phoenician "Phoenix" which means date palm, and from the Greek word "dactylifera" which means a finger, due to the resemblance of the fruit precisely with a finger, however, the author points out that its name could also come from the Egyptian bird "Fenix" which lived until 500 years old and after being thrown into the fire emerged again with a renewed growth, similar to what happens with the date palm, capable of surviving the damage by fire and growing again.

Table 5 describes the date palm species and their geographical distribution in the world.

Table 5. Distribution of date palm (*Phoenix dactylifera L.)* species.

Species	Common name	Distribution
Phoenix dactylifera	Date palm	Mediterranean countries, Africa and part of Asia, introduced in North America and Australia.
P. atlantica A. Chev		East Africa and Canary Islands
P. canariensis chabeaud	Canary Island palm	Canary Islands and Cape Verde

P. reclinata Jacq.	Saw palmetto	Tropical Africa (Senegal and Uganda) and Yemen (Asia)
P. sylvestris Roxb.	Date palm	India and Pakistan
P. humilis Royle	Date palm	India, Burma and China
P. hanceana Naudin		South China and Thailand
P. robelinic O'Brien		Sri Lanka, Toukin, Annam, Laos and Thailand
P. farinifera Roxb	Pygmy palm	India, Ceylon and Annam

Source: CHEVALIER, A. (1952): Recherche sur les Phoenix africains; RBA, Mai-Juin, 1952. Cited by Abdelouahhab (2002 Op. Cit.), apart from the species *P. dactylifera*, the edible species are *P. atlantica Chev, P. reclinata Jacq, P. farinifera Roxb, P. humilis Royle, P. acaulis Roxb.*

Most of the twelve Phoenix species are well known as ornamentals, the most valued being P. canariensis Chabeaud, commonly called Canary Island Palm. P. sylvestris Roxb. It is widely used in India as a source of sugar. P. dactylifera L. is distinguished from the two previous species by several characteristics which can be summarised as follows: production of reeds, tall columnar and relatively tall trunk, if the crown of the leaves is included it can reach up to 20 m in height, in relation to the fruit, the date palm weighs from 2 to 60 grams, depending on the variety, the water content of the fruit ranges from 24 to 85% depending on the variety and whether it is early or late. The extreme limits of the date palm's geographical distribution are between 10° North (in SomaKa) and 39° North (Spain), but between 24° North and 34° North (Morocco, Tunisia, Algeria, Israel, Egypt, Iraq, Iran) are the most favourable areas. As for the height above sea level, the date palm grows from 6 m above sea level (as in India, California,

U.S.A.) to 1500 m above sea level.As for the sugar content, it depends on the variety and percentage of humidity. For the USA, it is estimated that fifty-one varieties have 77% sugar content, and for sucrose alone, the percentages are 1 and 36% respectively for fresh and dry date palms (Abdelouahhab, 2002 Op. Cit).

The same source above indicates that some of the products derived from the fruit are juice, syrup, low calorie sweetener kquid sugar, protein yeast, vinegar, wine and alcohol, not counting their use as sweets and snacks, or as confectionery by coating them with chocolate, or nut fillings, jams, date butter, date cream, preserves and caramelised date palm, and it is determined that total sugars in high moisture content date palm average 78%.

According to Table 6, it is estimated that there are 100 million date palms in the world, distributed over 770 thousand ha. Iraq is the country with the highest concentration of date palms, with 22.3 million palms, contributing 22.3% of the world total of palms, Mexico only contributes marginally, as it is included in what Abdelouahhab (2002 Op. Cit) calls "Other countries" in his table.

Table 6. Area and total number of date palms worldwide.

Country	Number of palms (thousands)	% of world total	Area (thousands of ha)	Planting density (palms/ha)
Iraq	22,300	22.3	125	178
Com	21,000	21	180	116
Saudi Arabia	12,000	12	45	148
Algeria	9,000	9	45	200
Egypt	7,000	7	45	200
Libya	7,000	7	27.5	254

Pakistan	4,375	4.37		
Morocco	4,250	4.25	84.5	50
Tunisia	3,000	3	22.5	133
Sudan	1,333	1.33		
Mauritania	1,000	1		
Oman	1,000	1		
Yemen	800	0.8	6.4	125
USA	359	0.35	3.44	105
SomaKa	204	0.2	3.7	50
Bahrain	200	0.2	0.35	577
Israel	200	0.2	1.6	125
Palestine	60	0.06	0.25	200
Kuwait	38	0.03		
Syria	12	0.01		
Other countries	4,929	4.92		
WORLD TOTAL	100,000	100	770	173

Source: Djerabi (1995), "Mediterranean Choices" (1996), cited by Abdelouahhab (2002)

Regarding the amount of water demanded by the crop, Liebenberg and Abdelouahhab (2002) for Algeria, table 4 shows that depending on the region, the amount of water per hectare ranges from 10,368 to 34,190 m^3 ha^{-1} , equivalent, after dividing by the 10,000 m^2 of a ha, to irrigation spreads of 1,038 and 3,419 m respectively.

Table 7. Approximate water requirements of date palm in different regions of Algeria . Source:

Own elaboration, PJ Liebemberg and A Zaid (this is table 50 of their chapter VII Date of palm irrigation, quoted by Abdelouahhab Zaid, 2012. *Op. cit*).

Scientist (year)	Region	Number of trees per ha	Approximate needs (m^3 /ha/year)
Rolland (1894)	Sahara (Algeria)	130	34,190
Rose (1898)	Ziban (Algeria)	144	10,368

Jus (1900)	Oved Rhir (Algeria)	130	22,750
Wertheimer (1957)	Ziran (Algeria)	120	15,000

The range of the metric demand per hectare for regions of Algeria, one of the main datetil producing countries in the world, indicated in the previous paragraph coincides with the range of water demand in tables 7 and 8, for Algeria's figures with those of Comondu in traditional gravity irrigation, since these data range from 13,000 m^3 ha^{-1} (Morocco) to 36,000 m^3 ha^{-1} (California, U.S.A.). But it does not coincide with technified irrigation, where the volume of water used per hectare decreases to volumes of water equivalent to one fifth and even one tenth of the volume of water demanded in traditional gravity irrigation, as in the case of pump irrigation in San Luis Rfo Colorado, Sonora, where, with 2,600 m3 per hectare, the volume of water used per hectare decreases to volumes of water equivalent to one fifth and even one tenth of the volume of water demanded in traditional gravity irrigation.

Table 8: Irrigation of date palms worldwide.

Site	Water quantity (m^3 per ha)
Algeria	15,000-35,000
California, U.S.A.	27,000-36,000
Egypt	22,300
India	22,000-25,000
Iraq	15,000-20,000
Jordan Valley, Israel	25,000-32,000
Morocco	13,000-20,000
South Africa	25,000
Tunisia	23,600
Comondu, BCS Traditional Gravity Irrigation	24,000-27,000

Comondu, BCS Pump Irrigation	4,800-5,400
San Luis Rios Colorado, Sonora drip irrigation	2,600

Source for table 8: Own elaboration, with based in Liebenberg and Abdelouahhab (2002) for data from Algeria to Tunisia, and Cisneros (1983) for traditional gravity-irrigated date grass with rolled water, and pumping in Comondu, BCS, Mexico and from FIRA (2019) for the San Luis Rios Colorado, Sonora date field irrigated by groundwater drip irrigation.

m^3 ha^{-1} , implies an irrigation rate of 0.26 cm, equivalent to only one tenth of the irrigation rate of between 2.4 and 2.7 m for traditionally irrigated dates with Comondu's runoff water, or even more, it would represent only 7% of the rate of water used per hectare in the production of traditionally irrigated dates in Algeria or the USA.

The main varieties of date palm (*Phoenix dactylifera L.*) reported by Cisneros (1983)[61] are:

a) Deglet Noor, late variety, producing 90 to 130 kg per plant per year.

b) Khadrawy, precocious and 50 to 70 kg per plant per year.

c) Zahidi, intermediate variety, yielding 90 to 120 kg per plant per year.

d) Kustawy, intermediate variety, with 70 to 90 kg per plant per year.

e) Halawy, early variety, with 60 to 70 kg per plant per year.

SFA-SAGARPA (2010)[62] adds the Medjool variety to the above, pointing out that Californian producers call this variety "the Cadillac of dates", since "it is undoubtedly the best of all and the one that fetches the highest prices,

---

[61]**Cisneros, Arias Rodrigo, 1983**. Observations of date cultivation in the region of San Ignacio, La Purisima and Comondu, Baja California Sur. Professional thesis. School of Agriculture, University of Guadalajara, Mexico.
[62]**Secretary of Agricultural Development-SAGARPA, 2010**. Statistical study on date production in the municipality of Mexicali, Baja California. P.15

although due to the type of sugars it contains, its fruit is more delicate and its conservation requires refrigeration. For this reason, it is recommended to market it before three months. Its dates are extremely long, with a soft and sweet pulp, of a brown colour with variable tonalities depending on the cultivation area and with a weight per unit of between 15 and 23 grams. It originates from Morocco", while the Deglet Noor variety has a shelf life of up to one year without refrigeration and the weight of its fruits ranges from 8 to 11 grams, the date of the Halawy variety is between 7 to 10 grams, with soft pulp, tender and sweet taste, and elongated shape, it is classified as one of the best varieties, while the Zahidi variety is known as "golden date" because of its light colour, rounded shape, medium size and not as sweet as the previous varieties, so it is consumed in preparations that require cooking, typical of Arab cuisine and pastry".[63]

---

[63] **Secretarana de Fomento Agropecuario-SAGARPA, 2010,** *Op. Cit.*

IV. MATERIALS AND METHODS
IV.1. Location of the study area

The DDR Comondu, is located in the municipality of the same name, Comondu, in Baja California Sur, its extreme coordinates are 23° 35' 25" to 26 ° 24' 16" north latitude and 110° 52' 07" to 112° 47' 11" west longitude, It borders to the North with the municipality of Mulege, BCS, to the South with the municipality of La Paz, BCS, to the East with the municipality of Loreto, and to the West with the Pacific Ocean, its capital is Ciudad Constitucion, its territorial extension is 12,547.3 km^2 , with an average altitude above sea level of 456 meters, 70,816 inhabitants in 2010 according to INEGI in its latest General Census of Population and Housing, its climate is mostly desert, the average annual temperature is just above 22 ° C, with 100 mm of rainfall per year on average (see figure 5).

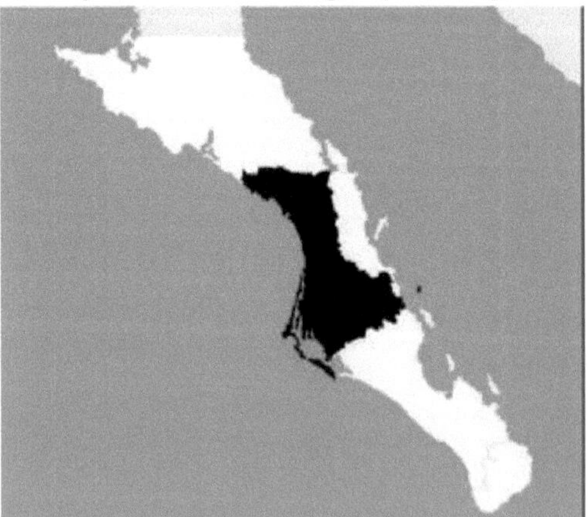

Figure 5. Location of the municipality of Comondu, Baja California Sur, Mexico.

IV.2. Sources of information

For the estimation of the physical, economic and social productivity indices of the cultivation of date palm (*Phoenix dactylifera) and* grain maize (Zea mays) in the DDR[64] Comondu, BCS, we used the variables of harvested area (in hectares), annual physical production (in tons) and the Gross Value of Production (in thousands of nominal pesos) reported by SIAP-SAGDER (2018) on their web page, on their web page, Annual physical production (in tons) and the Gross Value of Production (in thousands of nominal pesos) reported by SIAP-SAGDER (2018)[65] in its web page, from which the variables of physical yield "RF" per hectare (in tons per hectare) and the PMR (average rural price, in nominal MX$ per hectare), as well as the production costs per hectare reported by FIRA (for grain maize) on its AGROCOSTOS webpage[66] , this source of production costs also contains information on the number of labourers per hectare, financial cost and land rent, (which were agglutinated into variable costs and fixed costs), and the cost for depreciation of machinery and equipment was added, as well as amortisation of the initial capital invested. The production costs per hectare for 2018 for date palm (*Phoenix dactylifera*) were obtained from Vega (2014)[67] , who, in the proforma statement of the business plan, makes a projection of income and costs for the five years from 2015 to 2019, taking as a basis both the growth rates of date palm prices and production costs as well as the

---

[64]The SGADER (Secretaria de Agricultura, Ganaderia y Desarrollo Rural) divides the country into DRs (Distritos de Riego), and the DRs are further divided into DDRs (Distritos de Desarrollo Rural) which are further divided into CADERs (Centros de Apoyo al Desarrollo Rural).

65SIAP-SAGDER (2018). CierreAgricola2018 . Available in: http://infosiap.siap.gob.mx/aagrIcola_siap_gb/icultivo/

[66]**FIRA, 2019.** AGROCOSTOS. Disponible en: https://www.fira.gob.mx/InfEspDtoXML/TemasUsuario.jsp

[67]**Vega, C. Brenda B. 2014**. Business plan for the commercialisation of date from the oasis of San Miguel and San Jose de Comondu. Professional Thesis. Autonomous University of Baja California Sur. La Paz, Baja California Sur, Mexico.

inflation rate, and are broken down into fixed and variable costs, to which are added the costs for depreciation of machinery and equipment as well as capital amortisation based on the initial investment in the establishment of the crop, thereby standardising the production costs per hectare for both datetil and grain maize crops in the Comondu RDA.

IV.3. Database and Variables

The determination of physical (PFA), economic (PEA) and social (PSA) water productivity, evaluated through the determination of indicators in this study for date palm (*Phoenix dactylifera*) and grain maize (*Zea mays*) crops in the Comondu RDD, was based on the indices for PFA, PEA and PSA of Rios *et al* (2018[68]) specific to an individual crop or a conglomerate of diverse crops Rios *et al.* (2015 to[69]), which are shown in Table 9, using the independent variables of a macroeconomic-agricultural nature, their significance is noted below:

RF_i = Physical yield of the i-th crop (in ton ha)$^{-1}$.

LR_i = Irrigation line of the i-th crop (in m).

EC_i = Efficiency of hydraulic conduction of the i-th crop. $0 < EC < 1$[70] .

S_i = Harvested area of the i-th crop (in ha).

p_i = Price of the product of the i-th crop (in MX\$ ton)$^{-1}$.

PC = Exchange rate, Mexican pesos (MX\$) per USD.

C_i = Cost of production per hectare of the i-th crop (in MX\$)$^{-1}$

$g_i = U_i$ = Profit per hectare of the i-th crop (in US\$ ha^{-1}

[68]Rios Flores, Jose Luis, *et al (2018, Op. Cit.)*
69Rios-Flores, Jose Luis *et al* (2015, *op cit*).

[70]If the source of water supply is close to the plot of the crop that will use the water, as well as in the case of a highly technical type of irrigation, EC will tend to approach 1, i.e. it will tend to have 100% efficiency, but if the source of water supply is far away from the plot or if it is a low technical type of irrigation, then EC will tend to retreat from 1 and approach 0.

)=RFi(pi/PC)-(Ci/PC).

Ji = Number of days invested per hectare in the i-th crop.

i = i-th crop under a particular form of irrigation (pumping, gravity).

288 = Number of working days per year per worker = 6 working days per week, times 48 weeks per year.

Table 9. Mathematical models used for the measurement of physical (PFA), economic (PEA) and social (PES) productivity of water used in production.

Variable	Model for individual cultivation	Model for a crop group aggregate
1) PFA (en L kg^{-1})	$Y = 10^4 * LRi * (RFi * ECi)^{-1}$	$y = \dfrac{10^4 \sum_{i=1}^{n} S_i\, LR_i\, (EC_i)^{-1}}{\sum_{i=1}^{n} S_i\, RF_i}$
2) EFA (kg m^3)	$Y = 10^{-1} * RFi * ECi * LRi^{-1}$	$y = \dfrac{10^{-1} \sum_{i=1}^{n} S_i\, RF_i}{\sum_{i=1}^{n} S_i\, LR_i\, (EC_i)^{-1}}$
3) EEA (m^3 USD de ganancia^{-1})	$y = \dfrac{10^4 \left(\dfrac{LR_i}{EC_i}\right)}{RF_i \left(\dfrac{P_i}{PC}\right) - \left(\dfrac{C_i}{PC}\right)}$	$y = \dfrac{10^4 \sum_{i=1}^{n} S_i\, LR_i\, (EC_i)^{-1}}{\sum_{i=1}^{n} S_i\, ((RF_i\, P_i - C_i)/PC)}$
4) PEA (miles de USD de ganancia hm$^-$^{3})	$y = 10^2\, g_i EC_i (LR_i)^{-1}$	$y = \dfrac{10^2 \sum_{i=1}^{n} S_i\, g_i}{\sum_{i=1}^{n} S_i\, LR_i\, (EC_i)^{-1}}$
5) PSA (Empleos hm$^-$^{3})	$y = \dfrac{25\; J_i}{72\; (LR_i\,/\,EC_i)}$	$y = \dfrac{25\; \sum_{i=1}^{n} S_i J_i}{72\; \sum_{i=1}^{n} S_i\, (LR_i\,/\,EC_i)}$
6) ESA (m^3 empleo^{-1})	$y = \dfrac{2.88 \cdot 10^6\, LR_i}{J_i EC_i}$	$y = \left(2.88 \cdot 10^6\right) \dfrac{\sum_{i=1}^{n} S_i (LR_i\,/\,EC_i)}{\sum_{i=1}^{n} S_i J_i}$

Source: Own elaboration, based on the mathematical models estimating PFA, PEA, PSA, EFA, EEA and ESA for an individual crop or for crop aggregates of Rios *et al* (2015 a) and Rios *et al* (2018).

In all the equations of the PSA, PEA, PSA, EFA, EEA and ESA, the volume "V" of water used per hectare by the crop was used to obtain the quotient formed by the irrigation sheet "LR" divided by the conduction efficiency "EC" of the hydraulic network (estimated at 90%) multiplied by 10,000, which is the number of square metres of one hectare. LR and EC are the usual ones in the region, which were obtained with the *Irrigation Programme* of the National

Centre for Disciplinary Research on Water Soil Soil Plant Atmosphere Relationship (CENID-RASPA), an agency of the National Institute of Forestry, Agriculture and Livestock Research (INIFAP-CENID-RASPA, 2006)[71] .

The capital productivity equations for this study were as shown in table 10.

Table 10. Mathematical models used for the measurement of social productivity of capital (SPC), social efficiency of capital (SCE), and Benefit-Cost Ratio (BCR/C).

Variable	Model
7) PSC (Jobs/USD invested)	PSC=(31250/9)*Ji*PCi/Ci
8) ESC (USD invested employment)$^{-1}$	ESC = 288*Ci/(Ji*PCi)
(9) RB/C (dimensionless)	RB/C = RFi*pi / Ci

Source: Own elaboration, based on the mathematical models estimators of the ESC, PSC and RB/C of Rios and Navarrete (2017).

The independent variables J, RF, p, C and PC are as indicated for table 4.

---

[71] **INIFAP-CENID-RASPA. (2006).** *Irrigation Programme.* [Accessed on: 1 September 2019]. Available at: https://cenidraspa.org/serg/serg_v1.php

V. RESULTS AND DISCUSSION.

V.1. Profitability, productivity indicators and efficiency of water used in the production of date palm (*Phoenix dactylifera*) traditionally irrigated by gravity versus grain corn (*Zea mays*) in Comondu, Baja California Sur in 2018.

According to figures from SIAP-SAGDER (2018)[72] the agricultural pattern of the DDR Comondu had 57 crops in 2018, which had an area of 26,457 ha harvested, which produced a value of MX$ 2,672.485 million, the date represented 0.25% of the area and district GVP, corn grain meanwhile occupied 17.39% of the harvested area and contributed 6.57% of the district GVP.

Table 11 shows that in the DDR Comondu, BCS, 65 ha of dates were harvested in 2018, with a production volume of 230 tons, which had a market value of USD 342.67 million, in relation to maize grain, date represented 1.4 of the harvested area of maize grain, and its VBP was equivalent to 4% of the size of the VBP of grammea, but the price/ton of date, at USD 1,489, was 7.5 times the price per tonne of maize grain.

The revenue per hectare of dates amounted to USD 5,271.8 ha^{-1} , 2.8 times the revenue of maize grain with USD 1,883.5 ha^{-1} .

Table 11. Area, production, production value, profitability and water use in the cultivation of datil (*Phoenix dactylifera*) irrigated in the traditional way by gravity versus ma^z (*Zea mays*) grain in ComondO, Baja California Sur in 2018.

72SIAP-SAGDER (2018). "Cierre agrfcola 2018". Available at https://nube.siap.gob.mx/cierreagricola/

Macroeconomic variables	(a) Date	(b) Grain maize	c=a/b
Harvested area (ha)	65.0	4,604	0.014
Production (ton)	230.0	$43,620.580	
VBP (thousands of MX$)	$6,670.0	$168,790.911	0.040
VBP (thousands of USD)	$342.67	$8,671.60	0.040
Yield (ton ha)$^{-1}$	3.538	9.474	
Price/tonne (USD)	$1,489.9	$198.8	7.5
income/ha (USD)	$5,271.8	$1,883.5	2.80
Fixed costs/ha (USD)	$355.4		
Variable costs/ha (USD)	$1,993.0		
Depreciation/ha (USD)	$64.2		
Total cost/ha (USD)	$2,412.6	$1,827.66	1.32
Profit/ha (USD)	$2,859.24	$55.83	51.2
# of day labour/ha	37.92	7.2	5.27
Irrigation sheet "LR", in metres.	2.4	0.55	4.36
Hydraulic driving efficiency	0.70	0.900	
Net volume of water used per ha	34,286m^3	6,111m^3	5.6
Volume of water used over the whole harvested area (hm)^3	2.23	28.14	0.08
Number of daily wages generated on the whole harvested area	2,464.8	33,148.8	0.07
Equivalent jobs generated on all harvested area	8.56	115.10	0.07
Capital investment in all harvested area (Millions of USD)	$0.157	$8.415	0.02

Source: Own elaboration, based on figures from Vega (2014) for production costs as of 2018 (projected by her to 2019 in business plan for San Jose Comondu, BCS) and for figures of harvested area, production and value of production for both crops based on figures from SIAP-SAGDER (2019) available at: https://nube.siap.gob.mx/cierreagricola/, for maize grain irrigation sheet from INIFAP-SAGARPA, 2017. "Agenda Tecnica Agricola Baja California Sur" and for dates the 2.4 m irrigation sheet from Cisneros, Arias Rodrigo, 1983. (p. 20).

In terms of production costs per hectare, production costs per hectare for dates amounted to USD 2,412.6 ha^{-1} , 32% higher than the USD 1,827.66 ha^{-1} cost of grain maize, which meant that dates had a mass profit per hectare of USD 2,859.24, 51.2 times the amount of profit of grain maize with USD 55.83 ha^{-1} (see table 11).

Table 11 shows that dates generate a large amount of work

in relation to grain maize, since the former requires an investment of 37.92 days ha^{-1} (equivalent to 303.36 hours ha^{-1} , equivalent to 0.135428571 jobs ha^{-1} , since it was established in Materials and methods, that 288 days per year is equivalent to one equivalent job), while grain maize requires only 7.2 days ha (equivalent to 57.6 hours ha , equivalent to 0.025 jobs ha 135428571), which means that the areas harvested of dates and grain maize generate a large amount of work.2 day wages ha^{-1} (equivalent to 57.6 hours ha^{-1} , equivalent to 0.025 jobs ha^{-1}), which implied that the harvested areas of date and grain maize generated employment for 8.56 jobs (equivalent to 2,464.8 day wages) and 115.10 jobs (equivalent to 33,148.8 day wages) respectively.

In relation to the water demand per hectare, having a high irrigation level, 2.4 m and a low efficiency (70%) of water conduction in the hydraulic network of canals[73] when irrigated by gravity-rolled water, the water requirement per hectare was 34,286 m^3 ha-1 in date, against 6,111 m^3 ha^{-1} in grain corn, which, being irrigated with underground water by pumping and having a higher efficiency of hydraulic conduction (90%), brought a favourable shadow price in its water consumption, implying in turn that the datil consumed 5.The water demand per hectare, when multiplied by the total harvested area of each crop, was 2.23 and 28.14 hm^3 ,

---

[73]The index of efficiency of hydraulic conduction (IECH) in the irrigation water conduction network is greater than zero and less than the unit, in non-technified irrigation the IECH is very small and tends to zero as it is less technified and there is a great distance between the water supply source for irrigation and the plot, For example, in the DR017 of La Comarca Lagunera, the IECH of the plots irrigated with water from the Lazaro Cardenas Dam, almost 220 km from San Pedro De Las Colonias, has an IECH of 0.44, that is, only 44% of the water released 220 km upstream reaches the plots, while in San Pedro De Las Colonias, an orchard that is supplied by a deep well 100 m from the plot will have an IECH close to 90% (=0.9) and if that water from the deep well is used in the form of pressurised irrigation by microcompuertas, the IECH will be around 93-95% (0.93-0.95).

i.e., although the unit water demand per hectare was much lower in grain maize, having a notorious extension of harvested area, it implied a higher water consumption (table 11).

Finally, from table 11 it can be seen that the capital investment was only USD 0.157 million in the date crop, equivalent to only 2% of the capital investment in the entire harvested area of maize grain, with USD 8.415 million.

Table 12 shows the different indicators of water productivity, as well as indicators of the productivity of the invested capital. In relation to the profitability of the crop, shown in the lower part of the table, it can be observed that the datefruit crop had a Benefit/Cost Ratio (BCR) of 2.19, which suggests that for every USD 1 invested in the production of this fruit, it was recovered and an additional surplus was obtained, in the form of profit before tax, of USD 1.19, which in relation to the average market rate, given by the CETES rate[74] [75] , equal to 7.43 percentage units, typifies the production of datetil as a highly profitable investment, especially in relation to the reference crop, maize grain, where the RB/C equals 1.03, which indicates that although the grammea was profitable, insofar as it made it possible to recover each USD invested, there was a profit of only 3 cents for each of those USD invested. Also, in relation to the market Kder rate, it suggests that it would have been preferable for the producers of maize grain to invest their money in the bank, and without any risk, it would have

---

[74]The CETES (Certificados de la Tesoreria de la Federacion) rate is the market Hder rate, which suggests the average profit percentage in Mexico, which in this case means that for every MX$1 invested, you get it back, as well as an additional surplus, in the form of profit, of 7.43 cents. As of 22 October, the 182-day CETES rate is 7.43%, available at:
https://www.banxico.org.mx/tipcamb/llenarTasasInteresAction.do?idioma=sp
[75]MUSD, initials for million US dollars

yielded a profit not of 3 but of 8%.

The other two indicators of invested capital are not of a financial nature like the RB/C, they are of a social nature, since they evaluate, on the one hand, the social productivity of capital index (PSC), the amount of employment generated by each MUSD75; thus, table 12 shows that in the crops of date and grain maize they had indicators of 54.6 and 13.7 MUSD jobs^{-1} , which would indicate that the same amount of investment, one MUSD, was able to produce four times (the index is 3.99) more jobs in the datetil crop than in the maize grain crop. There are two reasons for this, the first is that in the case of datetil each hectare requires 37.92 day labourers, while in the case of maize grain it is only 7.2 day labourers per hectare, a little more than 7.The second cause is the different amount of capital invested per hectare in each crop. It should be remembered that in the case of datetil, the cost per hectare (USD 2,412.6 ha^{-1}) was 32% higher than the cost per hectare of grain maize (USD 1,827.66 ha-1).

Table 12. Productivity and physical, economic and social efficiency of water used in the production of the Datil *(Phoenix dactylifera)* crop irrigated in a traditional way by gravity-rolled water vs. grain maize in Comondij, Baja California Sur, Mexico.

Variable Units	It reads:	(a) Date	(b) Grain maize	c=a/b
Efficiency (E) y physical (P) productivity (F) of water (A) used in production				
EFA $m^3\,kg^{-1}$	*physical efficiency* of the water used in production	9.69	0.65	15.02
PFA $kg\,nr^3$	*physical productivity* of water used in production	0.103	1.550	0.07
Efficiency (E) y economic (P) productivity (F) of water (A) used in production				
Water efficiency (E) y water economic (P) productivity (E) (A)				
EEA m^3 of water used per USD gain	*economic efficiency* of the water used in production	11.991	109.452	0.11
EAP USD profit per m^3 water used in the production	*economic productivity* of the water used in production	$ 0.083	$ 0.009	9.13
Efficiency (E) y social (P) productivity (S) of water (A) used in production				
ESA m^3 of water used for each job created	*social efficiency* of water used in production	260,398	244,444	1.07
PSA Jobs generated by hm^3	*social productivity* of water used in production	3.84	4.09	0.94
Price of . ₁₀ᵣ> , USD nr^3 **water**	*price per m^3* of water used in production	$ 0.008	$ 0.079	0.10
Efficiency (E) y social (P) productivity (S) of Capital (C) used in production				
ESC USD invested per job created	*social efficiency* of capital invested in production	$18,323.54	$73,106.33	0.25
PSC Jobs generated per Million USD invested	*social productivity* of capital invested in production	54.6	13.7	3.99
RB/C Dimensionless		2.19	1.03	2.12

Source: Elaboration own, with figures from table 1.
The social capital efficiency ratio (ESC) number equal to

USD 18,323.54 employment^{-1} from table 12, suggests that creating one job required a capital investment of USD 18,323.54, or seen another way, that was the cost of generating one new job in the date crop, equivalent to 25% of what it cost to create one job in the maize grain crop: USD 73,106.33.

This suggests then, that any amount of capital invested in the cultivation of datetil will generate four times more employment (the index number in Table 12 was 3.99), than in the cultivation of maize grain, since when capital investment is seen as a social productivity index, it yielded the figure of 54.6 jobs MUSD^{-1} , while in the cultivation of maize grain the indicator was equal to 13.7 jobs MUSD^{-1} .Although this positions date as a highly employment-generating crop, at least in relation to maize grain, which is an unquestionable social *benefit*, it also presupposes that this crop, by demanding a large amount of labour, has lagged behind in terms of mechanisation, so it has a relatively high rate of work per kg of product: 0.0857 hours per kg[76] , which in a way does not favour it, since any other date produced in the country or in the world is not a good example.

Another part of the world, which requires less labour per kg, would have a better chance to penetrate the market and thus position and displace the Comondu date, following the same procedure it is obtained that in the case of maize grain a total of 0.0060 h $_{kg-1}$ is required.

The RB/C of dates, 2.19, suggests a profit rate of 119%, well above the CETES rate, equal to 7.43%, which places it as a highly profitable crop, but not maize grain, where the

---

[76] Derived by dividing 303.36 labour hours per ha (equal to the product of 37.92 labour hours per ha times 8 hours) by the 3,538.46 kg of date per hectare of physical yield.

RB/C, equal to 1.03, places it as an unattractive crop, since, after facing the problems of production, climate and market, it only obtains a 3% profit, well below the CETES rate.

The physical productivity of water (PFA) in the datetil crop was 0.103 kg m^{-3} , 7% of the water used in the maize grain to produce the same kg of biomass, since in the grammea the indicator was equal to 1.550 kg m^{-3} ; which, seen as its inverse, gives the physical water efficiency indicator (EFA), which indicates the necessary volume of water, in m^3 , used in the production to produce one kilogram of product, which in the case of date was 9.69 m^3 kg^{-1} , 15.02 times the volume of water used by grammea to produce one kg of ma^z: 0.65 m^3 kg^{-1} (table 12).

From Table 12 it can be seen that the economic water productivity (EWP) yielded a mdice number equal to USD 0.083 m^{-3} in date grass and USD 0.009 m^{-3} in grammea, i.e. the date grass crop was 8.13 (= 9.13 -1) times more productive in economic terms when using water in production, in other words, the same volume of water, one m^3 , produced a pre-tax profit equal to 8.3 US cents in the production of the desert fruit tree, while that same volume of water only produced a profit of 0.9 US cents.

As a measure of the economic water efficiency (EEA) with which water was used in production, Table 12 shows that the EEA in datetil was 11,991 m^3 USD^{-1} , while grammea reported a rate of 109.452 m^3 USD^{-1} , i.e. producing a profit mass of one US dollar, implied the water investment in datil of only 11% of the water invested in grain maize with almost 110 m^3 , suggesting that datil was more economically efficient in using water in production.

The social water productivity (SWP) of datetil was 3.84 jobs hm^{-3} , while in one of the main crops of the Comondu DDR, ma^z grain, the same volume of water, one million m^3 , produced 4.09 jobs, suggesting that while the datetil crop was more productive when using water in economic terms (USD 0.083 m^{-3} versus USD 0.009 m^{-3} respectively datetil and ma^z grain), it was not more productive in physical terms (0.103 kg m^{-3} versus 1.550 kg m^{-3} respectively datetil and ma^z grain) nor in social terms (3.84 versus 4.09 jobs hm^{-3} respectively datetil and ma^z grain), since datetil produced only 94% (the index in Table 12 was 0.94) of the level of employment produced by the same volume of water used in ma^z grain.

The inverse of the PES is the SWE (social efficiency of water) index, which shows in table 12 that in maize it was necessary to invest a volume of water equal to 260,398 m^3 to produce one job, while in grammea the volume of water was 244,444 m^3 , which shows that maize required a 7% higher volume of water (the index was 1.07 in table 12) than maize grain.

Table 12 shows the price of water, which although it seems to be an indicator of an economic rather than a social nature, insofar as it indicates how much water costs the agricultural producer[77] , it is in fact an indicator of an eminently social nature, since water is an extremely limited resource and, according to the Constitution of the United Mexican States, belongs to society as a whole, it is therefore a social resource, but one that is used privately and on the basis of which there is a de facto appropriation,

---

[77] The water price indicated in table 12 comes, as indicated in the methodology part, from dividing the amount of the irrigation cost item (within the costs per hectare) by the volume of water used per hectare.

[78]**Seckler, D., Upali, M; Molden, D.; De Silva, R.; Barker, R. 1998.** World water demand and supply, 1990 to 2025: Scenarios and Issues. Research Report 19. International Management Institute: Colombo, Sri Lanka. 40p. Available at: http://protosh2o.act.be/VIRTUALE-BIB/Water_in_de_Wereld/ALG-Algemeen/W_ALG_E22_World_Water.PDF/ (last accessed September 18,2019)

According to the Constitution of the United Mexican States, water belongs to society as a whole, it is therefore a social resource, but it is used privately and, on the basis of this, there is a *de facto* private appropriation of profits that have been generated with a resource that does not belong to the producer, but belongs *de jure* to society as a whole. Thus, the water price indicator shows (in the case of the datetil producer with a price index equal to USD 0.008 m3) that the datetil producer invested 0.8 cents for each m^3 of water used in production, which was paid to SAGARPA (currently SAGDER) in the form of a fee paid to the Secretary of Agriculture for allowing him to use surface water irrigated by gravity, while the corn grain producer, who used underground water and irrigated his plot by gravity, cost him 7.9 cents for each m^3 of water used in production.

In the discussion part, this water price incurred by the Comondu date producers will be contrasted with water prices for other crops in other parts of the country and in other parts of the world.

V.2. Discussion: contrast with the efficiency and productivity
of water used in the production of other crops.

The productivity of water used in the production of dates in Comondu must now be relativised against the water productivity of other crops in the same region of Comondu, apart from maize grain, as well as against the AFP, ADP and WFP of other crops and even other products such as milk. This is due to the fact that we are not studying the cultivation of dates, what we are studying is the PFA, PEA and PSA, since according to Flores (1978) it is Economics that provides all the theory, the instruments and the methodology, agriculture is only the raw material on which the baggage provided by Agricultural Economics is based.

To make the discussion part easier, in which the EAP and PES results determined for the datetil crop in the Comondu, BCS RLD are contrasted, and as the EAP monetary figures were valued in monetary figures from different years and in different countries (in table 12), they were first deflated to constant 2018 USD, to make the contrast valid against the datetil EAP figures, then schematized in table 13. In order to make this part of the discussion easier, only the crop against which the Comondu date is being contrasted will be mentioned, which appears on the far left of Table 13, and the author who determined the data, which appears on the far right of Table 13, will not be mentioned.

In relation to economic water productivity (EWP), according to Seckler (1998)[78], this is nothing more than either the total value or the net value divided by the amount of water applied or allocated to production, and its importance lies in the fact that by knowing the EWP, the opportunity cost of the different alternative uses of water can be established, i.e. if water is used in activity "A" it cannot be used in activity "B", so the difference between the value generated between the two activities will act as an opportunity cost, if water is used in activity "A" it cannot be used in activity "B", so the difference between the value generated between both activities will act as an opportunity cost, since, for example, if EAP "A" minus EAP "B" is positive, then, this will mean that allocating water to the production of "B" will imply not obtaining an income equal to the difference between EAP "A" and EAP "B".

From Table 13 it can be seen that the Comondu DDR datil, with an ADP index of USD 83,395 hm-3 was less economically productive in terms of water use in production than the orange crops produced there in Comondu, the apple crop produced under low, medium and high technology conditions, the grape crops produced in Caborca, Sonora and Coahuila, the peach crop produced in Fresnillo, Zacatecas as well as all the agricultural products listed in the table for Spain: cotton, rice, strawberry, sunflower, grain corn and olive. It should be noted that the average EAP found was USD 1,435, 930 hm^{-3} , 17.22 times the level of the EAP of the date.

Table 13 shows that the highest EAP in Mexico was recorded for the grape crop produced in Caborca, Sonora (with USD 1,012,511 profit per hm^{3}), 12.14 times the size of the index of USD 83,395 hm^{-3} of datetil, and abroad, in Spain, strawberry cultivation with USD 25,987,394 hm^{-3} , had an EAP 311.62 times the size of the index of the EAP of datetil (USD 83,395 hm)$.^{-3}$

Relative economic (PEA) and social productivity (PES) of water used in the production of various perennial and annual agricultural and livestock products. Comondd's date =1.

Product	Place	Standardised ADP at constant 2018 USD per hm^3	EAP Datil= 1	PSA	Units	PSA Datil= 1	Author
Datil	Comondu, BCS	$ 83,395	1.00	3.84	Jobs hm^3	1.00	This work
Maize grain	Comondu, BCS	$9,136	0.11	4.09	Jobs hm^3	1.07	This work
Orange	Comondu, BCS	$90,916	1.09	3.60	Jobs hm^3	0.94	Cifuentes, 2013.
apple tree BT	Cuauhtemoc, Chihuahua	$ 90,348	1.08	28.10	Jobs hm^3	7.32	Rios et al (2017)
Apple AT	Cuauhtemoc, Chihuahua	$657,994	7.89	22.70	Jobs hm^3	5.91	Rios et al (2017)
Apple MT	Cuauhtemoc, Chihuahua	$305,006	3.66	19.60	Jobs hm^3	5.10	Azpilcueta et al (2017)
Temporary coffee	Villaflores, Chiapas	$ 24,630	0.30	26.30	Jobs hm^3	6.85	Azpilcueta et al (2017)
Temporary coffee	Chiapas	-$21,740	-0.26	16.30	Jobs hm^3	4.24	Azpilcueta et al (2017)
Vine	Caborca, Sonora	$1,012,511	12.14	10.70	Jobs hm^3	2.79	Rios et al (2018)
Vine	Coahuila	$239,922	2.88	53.32	Jobs hm^3	13.88	Carrillo, 2019
Avocado	Patzcuaro, Michoacan	$117,835	1.41	31.90	Jobs hm^3	8.31	Rios, Torres y Azpilcueta (2019)
Walnut	Delicias,	$ 59,543	0.71	3.90	Employments hm^3	1.02	

					Rios, Towers y Torres (2016)
Walnut	Southwest de	$ 34,179	0.41	17.10 Jobs hm³ 4.45	Rios and Navarrete (2017)
Peach	Fresnillo,	$113,675	1.36	2.00 Jobs hm³ 0.52	Rios et al (2015)
Milk bovine	Delicias, Chihuahua	$ 4,821	0.06		Rivers, Rivers and Rivers (2019)
Cotton	Spain	$279,304	3.35		Montesinos et al (2011)
Rice	Spain	$510,033	6.12	sd Sd	Montesinos et al (2011)
Strawberry	Spain	$25,987,394	311.62	sd Sd	Montesinos et al (2011)
Sunflower	Spain	$303,591	3.64	sd Sd	Montesinos et al (2011)
Maize grain	Spain	$510,033	6.12	sd Sd	Montesinos et al (2011)
Olive tree	Spain	$1,177,933	14.12	sd Sd	Montesinos et al (2011)
AVERAGE		$1,435,930	17.22	17.39 Sd	

Source: Own elaboration, based on tables 1 and 3.

Table 13 shows that the date crop in the Comondu DDR had a higher EAP than only the maize grain crops in the same DDR of Comondu, the coffee produced under rainfed conditions in Villaflores, Chiapas, as well as the average coffee produced in the entire state of Chiapas, the pecan walnut tree both in Chihuahua and in southwestern Coahuila, as well as the EAP of the specialised bovine milk produced in Delicias, Chihuahua, which in fact, reported the lowest EAP, with an index of USD 4,821 hm^{-3} , equivalent to only 6% (the index was 0.06) of the amount of profit generated by one hm^3 of water used in datetil production in Comondu, BCS.

In relation to PES, Table 13 shows that Comondu date, with

3.84 jobs hm^{-3} , was less socially productive in using water in production than the maize grain crops of the same DDR Comondu (with 7% higher PES), the BT, MT and AT apples produced in Cuauhtemoc, Chihuahua (whose PES indices were 7.32, 5.91 and 5.10 times the PES of date, coffee produced with rainwater in the DDR Villaflores, Chiapas (with PES indices of 7.32, 5.91 and 5.10 times the PES of date), coffee produced with rainwater in the DDR Villaflores, Chiapas, as well as at the statewide level of Chiapas (with PES indices 6.85 and 4.24 times the size of the PES index of date), grapevine produced in both DR037 Caborca, Sonora (with PES 179% higher than date) and grapevine at the statewide level of Coahuila (whose PES was almost 14 73% higher than that of date), and coffee produced with rainwater in the DDR Villaflores, Chiapas, as well as at the statewide level of Chiapas (with PES indices 6.85 and 4.24 times the size of the PES index of date).

times that of date), walnut produced in Chihuahua (with PSA 2% higher than date) and south-west Coahuila (with PSA 345% higher than date), finally, avocado produced in Patzcuaro, Michoacan had a PSA 731% higher than date.

The PES of date, with 3.49 jobs hm^{-3} , was superior only to the PES of peach produced in Fresnillo, Zacatecas (with PES 48% less than that of date), and orange produced in the same RDD Comondu, BCS whose PES was 6% less than date.

The price per m^{3} of water used in production is indicated in Table 14, and from that source we can observe that for nine crops/location, the average price per m^{3} (already deflated and valued in constant 2018 USD) of water was USD 0.043, noting also that the price of surface water used in the production of date, was the cheapest, with USD 0.008 m^{-3} , since the remaining crops, all, without exception, had a higher price of water, the crop with the lowest price of water was another fruit crop like date: peach produced in fresnillo, with USD 0.010 m^{-3} , only 27% more expensive water than for the Comondu date, while the crop with the highest price for water used in production was the vine irrigated by pumping with groundwater, since the price of the water used was 9.93 times what the Comondu farmer paid for the water used to irrigate his date palm, the same Comondu maize grain irrigated with groundwater had a price in water used equivalent to 9.73 times the price of the water used in the date palm, finally, the cost of the groundwater used to irrigate the avocado produced in Patzcuaro, Michoacan, cost 60% more than the water used in the Comondu date palm (see table 14).

Nunez *et al* (2000)[79][80] and Takele and Kallenbach (2001)[80], both in relation to water prices in alfalfa cultivation, point out that water prices are important in that they serve to improve demand as well as to influence the conservation of the water resource, however, the price per cubic metre of water is not in itself an indicator of the real value of water, for example, some farmers in the USA pay between USD 0.01 and USD 0.05 per cubic metre of water used in production, in irrigation to be exact, but the price of water paid for domestic consumption is not in itself an indicator of the real value of water.U.A. pay between USD 0.01 and USD 0.05 for each m^3 of water used in production, in irrigation to be

[79] **Nunez, H. G., Chew, Y. I., Reyes, J. I. 2000**. Production and utilization of alfalfa in northern Mexico. GHJ. Editors. Technical book. No. 2 SAGARPA-INIFAP.CIRNOC.CELALA. Matamoros, Coahuila, Mexico. 171 p.

[80] **Takele, E. & Kallenbach, R. 2001.** Analysis of the Impact of Alfalfa Forage Production under Summer Water-Limiting Circumstances on Productivity, Agricultural and Growers Returns and Plant Stand. Journal of Agronomy and Crop Science. 187(1): 41-46.

exact, however, the price of water paid for domestic consumption, according to Gleick, P. H. 2004 (*Op. Cit)*, the consumer pays from USD 0.30 to USD 0.80 m^{-3} , for water treated for domestic use, in the same sense, the Israeli farmer producing tomato (*Solanum lycopersicum*)
pays USD 0.57 m-3.

Table 14. Price of water used in the production of the date crop produced under gravity irrigation in the ComondiJ RDD, BCS, compared to other crops.

Crop/Location	Reported by the author		Price per m³ (standardised to constant 2018 USD)	Author	Datil=1
	Price per m³ (MX$ o USD nominal)	Units			
Datil, DDR Comondu	$ 0.008	Constant 2018 USD per m³	$ 0.008	This work	1.00
Maize Grain, DDR Comondu	$ 0.079	Constant 2018 USD per m³	$ 0.079	This work	9.73
Orange DDR Comondu, BCS	$ 1.110	MN$/m3	$ 0.074	Cifuentes, 2013.	9.21
Vine, caborca, Sonora	$ 0.070	USD/m3	$ 0.004	Rios *et al* (2018)	0.54
Vid, Coahuila	$ 0.075	USD/m3	$ 0.005	Carrillo, 2019	0.58
Avocado, Patzcuaro, Michoacan	$0.012	USD/m3	$ 0.001	Rios, Torres y Azpilcueta (2019)	0.09
Nogal, Delicias, Chihuahua	$ 0.550	MX$/m3	$ 0.037	Rios, Torres y Torres	4.56

				(2016)
Nogal, Southwest Coahuila	$0.016	USD/m3	$ 0.001	Rios y Navarrete 0.13 (2017)
Durazno, Fresnillo, Zacatecas	$0.162	MX$ per m3	$0.011	Rios et al 1.38 (2015)
AVERAGE			$ 0.024	3.02
Source:		Elaboration		own.

VI. CONCLUSIONS AND RECOMMENDATIONS
6.1. Conclusions

The particular objectives of determining the profitability, by means of the RB/C, were fulfilled, as well as the objective of determining numerical indicators on water efficiency and productivity in physical, economic and social terms of the water used in the production of the datil crop (*Phoenix dactylifera*) in the DDR Comondu, Baja California Sur and contracting them against the corresponding indicators of the grain maize crop in the same DDR.

Based on the methodology and mathematical models used, and the results obtained by such models, the first hypothesis is accepted, as the hypothesis was fulfilled: that the datil crop (*Phoenix dactylifera*) had a higher profitability, since its RB/C was 2.19, which in relation to maize grain, with a RB/C of 1.03, was 2.12 times higher.

Based on the methodology and mathematical models used and the results of such models, the second hypothesis is rejected, since the PFA rate of the datil (0.103 kg m^{-3}) was only 7% of the PFA of the grain maize (*Zea mays*) which produced 1.55 kg of biomass for each m^3 of water used in production. Based on the methodology and mathematical models used and the results of these models, the third hypothesis is accepted, since the ADP of the datil (*Phoenix dactylifera*) crop with USD 0.083 profit per m^3 of water used in production was 9.13 times greater than the ADP of the grain maize (*Zea mays)* crop in the DDR Comondu, BCS.

Based on the methodology and mathematical models used, and the results of these models, the fourth hypothesis is rejected, since, contrary to what was assumed, the cultivation of date (*Phoenix dactylifera)* had a lower PES rate than maize grain (*Zea mays),* which in practical terms, is indicated in this PES rate, that the water used in the production of date (*Phoenix dactylifera*), with 3.84 jobs hm^{-3}, produces 6% less jobs than the same volume of water used in the cultivation of grammea.

6.2. Recommendations

Since water is an extremely scarce resource and has multiple uses, the Agricultural-Environmental Economy must assign that scarce water to the alternative that maximises economic benefits such as profits or social benefits such as employment, while minimising the amount of water to be used to achieve that objective, since the greatest benefit is not the production of profits or employment, but the fact that water can be used now, In order for the environmental agricultural economy to achieve this, it must be based on the use of physical, economic and social water productivity indexes, which are suggested to be used by those institutions in charge of allocating water resources at the federal, state or municipal level among the different alternatives of use to which water is subject.

LITERATURE CITED

Abdelouahhab, Zaid, 2002. Date palm cultivation. FAO OLANT PRODUCTION AND PROTECTION PAPER. 156 REV.1 ISSN 0259-2517 ISBN 92-5-104863-0. Rome, Italy. Available at: http://www.fao.org/3/y4360e/y4360e00.htm

Azpilcueta, Ruiz-Esparza, M de Jesus; Rios-Flores, J. Luis; Ruiz-Torres, Jose. 2017. Water productivity in rainfed coffee crops in Villaflores, Chiapas, Mexico. Revista Asuntos Economicos y Administrativos No. 32, Primer semestre 2017. ISSN 0124-1133. University of Manizales, Colombia. pp. 183-192.

Agua.org.mx Fondo para la Comunicacion y la Educacion Ambiental A.C. Overview of water in Mexico, 2019 Available at: https://agua.org.mx/cuanta-agua-tiene-mexico/. Last accessed on: 10 September, 2019.

BANAMEX, 2018. Accessed on 24 September 2019. Mexican peso-US dollar exchange rate. Available at: https://www.banamex.com/economia_finanzas/es/divisas_ m etales/divisas_divisas_mundiales.htm

Carrillo, C., J. 2019. Economic-social productivity of water in drip-irrigated grapevine (*Vitis* vinifera) in Coahuila, Mexico. Professional thesis. Autonomous University Chapingo, Bermejillo, Durango, Mexico.

Chevalier, A. 1952. Recherche sur les *Phoenix* africains; RBA, May-June, 1952. Cited by Abdelouahhab (2002 *Op. Cit.*).

Cifuentes, G. O. 2013. Physical, economic and social efficiency of irrigation water in orange (*Citrus sinensis*) cultivation in Baja California Sur. Professional thesis. Universidad Autonoma Agraria Antonio Narro-Unidad Laguna, Torreon, Coahuila.

Cisneros, Arias Rodrigo, 1983. Observations of date

cultivation in the region of San Ignacio, La Purisima and Comondu, Baja California Sur. Professional thesis. School of Agriculture, University of Guadalajara, Mexico.

DRANSFIELD, J. and NW UHL. (1986): A summary of a classification of palms. Principles 30 (1): 3-11. Cited by **Abdelouahhab, Zaid, 2002**.

INIFAP-CENID-RASPA. (2006). *Programa Riego.* [Accessed on: 01 May 2017]. Available at: https://cenidraspa.org/serg/serg_v1.php

Cosmo News, 2012. ^How much water is there on earth? Available from: https://www.cosmonoticias.org/cuanta-agua-hay-en-la-terra/, Accessed 10 September, 2019.

Nacional del Agua, (2015): *Atlas del Agua en Mexico.* Conagua. Document available
at: http://www.conagua.gob.mx/CONAGUA07/Publicacione s/Publications/ATLAS2.

El Heraldo, Economy Section. 22 March 2015. Costa Rica. Available at:
https://www.elheraldo.co/economia/la-agricultura-consume-el-70-of-the-water-in-the-world-188535#

Evolution of the world population. The market economy: virtues and innovations. Demographics, 2019. Available in:
http://www.juntadeandalucia.es/averroes/centros-tic/14002996/helvia/aula/archivos/repositorio/250/271/html/e conomia/2/evolucion.htm. Accessed 10 September 2019.

FAO, 2002. FAO, undated. Available at:
http://www.fao.org/3/y3918s/y3918s03.htm *Water and crops. Achieving optimal water use in agriculture.* Food and Agriculture Organization of the United Nations: Rome, Italy.

FIRA, 2019. [On line]. *AGROCOSTOS* [Accessed 20 September 2019]. Available at:

https://www.fira.gob.mx/InfEspDtoXML/TemasUsuario.jsp

Fruits and vegetables. Available at: https://www.frutas-hortalizas.com/Frutas/Origen-produccion-Datil.html

Aquae Foundation, 2019. Quantity of drinking water, source of life. Available at: https://www.fundacionaquae.org/wiki-aquae/datos-datos-del-agua/cantidad-de-agua-potable-source-of-life/ date accessed: 10 September, 2019.

Gleick, P. H. 2004. Global freshwater resources: Soft-path solutions for the 21st century. Science. 302: 1524-1528.

Hoekstra, A.Y. 2003. Virtual Water Trade: Proceedings of the International Expert Meeting on Virtual Water Trade. Delft. The Netherlands. 12 and 13 December 2002. Value of Water Research Report Series No. 12. UNESCO-IHE. Delft. The Netherlands.
www.waterfootprint.org/Reports/Report12.pdf

Hoekstra A.Y., and Chapagain A.K. 2004. Water Footprints of Nations. UNESCO-IHE. Institute for Water Education. Value of Water. Research Report Series. Series 16. Volume 1. Netherlands.

Infoagro. Date cultivation (part 1): https://www.infoagro.com/frutas/frutas_tropicales/datil.htm

INIFAP-CENID-RASPA. (2006). *Irrigation Programme.* [Accessed on: 1 September 2019]. Available at: https://cenidraspa.org/serg/serg_v1.php

Kijne, J.W., R. Barker and D. Molden, 2003. Water Productivity in Agriculture: Limits and Opportunity for Improvement. CABI, Cambridge, UK, ISBN: 0 85199 669 8.

Liebemberg, P. J., Abdelouahhab, Zaid. 2002. Date of palm irrigation. Cited by Abdelaouahhab, Z. 2002.

Mekonnen, M.M. & Hoekstra, A. G. (2012). A global assesment of the water footprint of farm animal products.

ECOSYSTEM (2012). 15:401-415. DOI:10.1007-s10021-011-9517-8.

Molden, D; Murray-Rust, H.; Sakthivadiel, R; Makin, I.; 2003. A water productivity framework for understanding and action, pp.1-18. In: Kijne, J. W.; Barker, R.; Molden, D. J. 2003. Water productivity in agriculture: limits and opportunities for improvement. CABI Publication, Wallingford UK. 332p.

Montesinos, P.; Camacho, E.; Campos, B.; Rodriguez-Diaz, J. 2011. Analysis of Virtual Irrigation Water. Application to Water Resources Management in a Mediterranean River Basin. Water Resources Management. 25 (6): 1635-1651. Cited by Rios *et al* 2018.

Nunez, H. G., Chew, Y. I., Reyes, J. I. 2000. Production and utilization of alfalfa in northern Mexico. GHJ. Editors. Technical book. No. 2 SAGARPA-INIFAP.CIRNOC.CELALA. Matamoros, Coahuila, Mexico. 171 p.

Rios- Flores, J. Luis, Torres M., Miriam, Castro F., Rafael, Torres M., M.A. Ruiz T. Jose. 2015 a. Determination of the blue metric footprint in forage crops of DR017 Comarca Lagunera, Mexico. Rev. FCA UNCUYO, 2018. 47(1): 101-122, ISSN print 0370-4661. ISSN (online) 1853-8665, pp.93-107. Mendoza, Argentina.

Rios-Flores, Jose Luis, Torres M. M. and Torres M., M. A. (2016 a). Agricultural water productivity of pecan nut trees in northern Mexico. Cases: Comarca Lagunera and Delicias, Chihuahua. ISBN978-3-639-80166-8. Editorial Academica Espanola. Saarbrucken, Germany.

Rlos-Flores, J.Luis and Navarrete_Molina, C. (2017). The monetary footprint and economic productivity of water in pecan walnut (Carya illinoensis*)* in the southwest of

Coahuila, Mexico. Journal: Estudios de Economia Aplicada. Volume 35-3, September 2017. ISSN 1133-3197. International Association of Applied Economics (ASEPELT), Spain.

Rfos-Flores, Jose Luis, Rios Arredondo, Becky Elizabeth, CantO Brito, JesOs Enrique, Rios Arredodndo, Hebrian Efrain, Armendariz Erives, Sigifredo, Chavez Rivero, Jose Antonio, Navarrete Molina, Cayetano & Castro Franco, Rafael (2018). Analisis de la eficiencia fsica, economica y social del agua en esparrago (*Asparagus officinalis L.*) y uva (*Vitis* vimfera) mesa del DR-037 Altar- Pitiquito-Caborca, Sonora, Mexico 2018. *Revista de la Facultad de Ciencias Agrarias. National University of Cuyo*, 50(2). ISSN in print 0370-4661, ISSN (online) 1853-8665. Mendoza, Argentina.

Rios-Flores, J. L., Rios A., Becky E., Rios A., Hebrian E. 2019. Physical and economic footprints of milk. The case of bovine milk from Delicias, Chihuahua, Mexico. Editorial Academica Espanola. ISBN 978-620-0-02518-0. Beau Bassin, Republic of Mauritius.

Rios-Flores, Jose Luis, Ruiz-Torres, J. Azpilcueta Ruiz-Espazra, M. 2019. Economic-social productivity of water in avocado cultivation. The case of production in Michoacan, Mexico. Editorial Academico Espanola. ISBN 978-3-639-53183-1. Beau Bassin, Mauritius.

Salomon-Torres, R., Ortiz-Uribe, N., & Villa-Angulo, R. 2017. The production of date palm (*Phoenix dactylifera L.*) in Mexico. Nueva epoca. Ano 16 No.91, January-June 2017, Ciencias Sociales y Exactas. Autonomous University of Baja California.

Seckler, D., Upali, M; Molden, D.; De Silva, R.; Barker, R. 1998. World water demand and supply, 1990 to 2025:

Scenarios and Issues. Research Report 19. International Management Institute: Colombo, Sri Lanka. 40p. Available at: http://protosh2o.act.be/VIRTUALE-BIB/Water_in_de_Wereld/ALG-Algemeen/W_ALG_E22_World_Water.PDF/ (last accessed September 18, 2019).

SIAP-SAGARPA (2016). Cited by Salomon, Ortiz and Villa (2017, *Op. Cit.*).

SIAP-SAGDER (2018). Agricultural closure 2018. Available at:

http://infosiap.siap.gob.mx/aagricola_siap_gb/icultivo/

Takele, E. & Kallenbach, R. 2001. Analysis of the Impact of Alfalfa Forage Production under Summer Water-Limiting Circumstances on Productivity, Agricultural and Growers Returns and Plant Stand. Journal of Agronomy and Crop Science. 187(1): 41-46.

Vega, C. Brenda B. 2014. Business plan for the commercialisation of date from the oasis of San Miguel and San Jose de Comondu. Professional Thesis. Autonomous University of Baja California Sur. La Paz, Baja California Sur, Mexico.

Viets, F.G. 1966. Increasing water use efficiency by soil management. In Plant environment and efficient water use. Guilford RD, Madison, USA: American Society of Agronomy. Soil science Society of America. 295 p.

Zaid, A. 2002. Botanical description of the date palm, In: Cultivation of date palms. FAO plant production and protection document. ISSN 0259-2517 ISBN 92-5-104863-0. Rome, Italy.

yes

I want morebooks!

Buy your books fast and straightforward online - at one of world's fastest growing online book stores! Environmentally sound due to Print-on-Demand technologies.

Buy your books online at
www.morebooks.shop

Kaufen Sie Ihre Bücher schnell und unkompliziert online – auf einer der am schnellsten wachsenden Buchhandelsplattformen weltweit! Dank Print-On-Demand umwelt- und ressourcenschonend produziert.

Bücher schneller online kaufen
www.morebooks.shop

info@omniscriptum.com
www.omniscriptum.com

Printed by Books on Demand GmbH, Norderstedt / Germany